# ANCIENT METHODS TO SEARCH GROUND WATER

## D S PANDEY

INDIA · SINGAPORE · MALAYSIA

# Notion Press

Old No. 38, New No. 6
McNichols Road, Chetpet
Chennai - 600 031

First Published by Notion Press 2019
Copyright © D S Pandey 2019
All Rights Reserved.

ISBN  978-1-68466-270-8

# CONTENTS

# PREFACE

This book entitled, "Ancient Methods to search the Ground water with scientific evaluation has been written mainly for ground water Scientist, Engineers, NGOS and the students striving to acquire some back ground and perspective on this subject. Emphasis have been given to correlate ancient methods based on Varahmihir Samhita – its hindi/English translation with scientific correlation with modern concept.

As water is a precious mineral and due to increase of population, industry and agriculture its growth have been increased in many fold so the ground water research ancient and its correlation with modern science is important for deliniation of good ground water promising zone.

This book is expected to be useful not only to university students, technologists, particularly ground water scientists, Drillers but also for all those who are practicing in this field.

This is the first attempt to bring such information among the ground water experts in this world. Definitely this book will be a guideline/pathfinder with some new methods approach to solve the ground water problems of this world.

# FOREWORD

The book ancient method to search the ground water with its scientific evaluation has been central theme of the human use of ground water for their survival is the serious challenge in the present – scenario. An attempt have been made in the simple way to find the ancient methods with its present scientifically investigated research in this field.

This book provides an indepth survey of ancient methods to search the ground water with its scientific evaluation and finding facts on ground.

I think this is a good attempt to have brief account on history of the ancient methods its significance and finding the facts in the present scenario for ground water exploitation and quality problems.

I hope that this publication will provide a path for ground water users, students and the professionals working in this field.

**Dr. R.N. Tiwari**
Former Professor and Head
Benares Hindu University Varanasi

# CHAPTER 1

Ancient method to search the Ground water with
scientific evaluation.

धर्म्यं यशस्यं च वदाम्यतोऽहं दकार्गलं येन जलोपलब्धिः ।
पुंसां यथाङ्गेषु शिरास्तथैव क्षितावपि प्रोन्नतनिम्नसंस्थाः ।। 1 ।।
एकेन वर्णन रसेन चाम्भश्च्युतं नभस्तो वसुधाविशेषात् ।
नानारसत्वं बहुवर्णतां च गतं परीक्ष्यं क्षितितुल्यमेव ।। 2 ।।

**हिंदी:** मैं वराह मिहिराचार्य धर्म यश का देनेवाला जिससे जल प्राप्त हो
ऐसा दकार्गल कहता हूँ। जैसे पुरुषों के शरीर में उपर नीचे शिरा रहती
है, उसी तरह पृथ्वी में भी जल वाहिनीशिरा रहती है। आकाश से जल
एकवर्ण एवं एक रस में पृथ्वी पर गिरता है परन्तु नानाप्रकार की भूमि
होने से अनेक रंग एवं स्वाद का होता है। इससे भूमि के समान जल का
रूप और गुण जाने।

**English:** I Varahmihir Acharya who offers praise and religion speaks about
the methods by which water can be obtained. As the blood circulating in
human body in the same way the water flow in the veins under the earth.

The water sprinkles from the sky in natural form and good quality
but the quality varies when it comes in contact of different type of soil
formations so there is a direct relationship between quality of water and
soil formation.

**Scientific Correction:**

It gives the theory of ancient concept with the present scientific concept.
Different aquifer zone I, II, III and IV etc. of sub surface.

The water which falls from sky is fresh and it taste differs depending upon the water coming in contact of different depth zones of sub surface formation.

पुरुहूतानलयमनिर्ऋतिवरूणपवनेन्दुशङ्करा देवाः ।
विज्ञातव्याः क्रमशः प्राच्याद्यानां दिशां पतयः ॥ 3 ॥
दिक्पतिसंज्ञा च शिरा नवमी मध्ये महाशिरानाम्नी ।
सताम्योऽन्याः शतशो विनिःसृता नामभिः प्रथिताः ॥ 4 ॥
पातालदूर्ध्वशिरा शुभा चतुर्दिक्षु संस्थिता याश्च ।
कोणदिगुत्या न शुभाः शिरानिमित्तान्यतो वक्ष्ये ॥ 5 ॥

**हिंदी:** इंद्र, अग्नि, यम, निर्ऋति, वरूण, वायु सोम महादेव यह आठ देवता पूर्वादिक आठों दिशाओं के क्रम से मालिक होते हैं जिस दिशा से जो शिरा (जल वाहिनी नस) आवे उसको उस दिशा के स्वामी के नाम से जाने अर्थात उसका नाम उसके स्वामी के नाम से कहा जायेगा।

पाताल (नीचे) से उपर जो शिरा आवे वह पूर्वादिक चारो दिशा की शिरायें शुभ हैं। अग्नि आदि चारो कोण की शिरायें शुभ को देने वाली नहीं हैं इनके अन्तर्गत चिन्हों को शिरा कहते हैं।

**English:** Indra, Yam, Nriti, Varun, Air, Some, Shankara these eight gods are the lords of eight direction (N, S, E, W, NE, SE, NW, SW). The water flowing veins which flow from a particular direction that should be named by the name of the same direction. In the centre, there is a large vein by which numerous veins flows.

A vein which flows from sub surface (greater depth) to surface and the other vein which flows from eastern direction they are ground water repositories. The other veins which flows from Agni direction (SE) are not so useful from ground water point of view.

**Scientific Interpretation:**

Deeper aquifer is a good source of ground water. Hence the vein or aquifer which flows from deeper side to upper side is a good ground water bearing zone.

यदि वेतसोऽम्बुरहिते देशे हस्तैस्त्रिभिस्ततः पश्चात् ।
सार्धं पुरुषे तोयं वहति शिरा पश्चिमा तत्र ॥ 6 ॥
चिह्नमपि चार्धपुरुषे मण्डूकः पाण्डुरोऽथ मृत् पीता ।
पुटभेदकश्च तस्मिन् पाषाणो भवति तोयमधः ॥ 7 ॥

**हिंदी:** जल रहित देश में यदि बेत का पेड़ हो तो उसके पश्चिम तीन हाथ पर डेढ़ पुरुष (9' या 2.70 मीटर) के नीचे जल रहता है और पश्चिम तरफ से सोती बहती है। पहले दिखाई पड़ने को सोती या शिरा कहते हैं।

आधा पुरुष (2.5 हाथ या करीब 1 मीटर के नीचे सफेद मेंढक दिखाई देगा तदन्तर पीली मिट्टी उसके बाद एक फुट (दोनों के समान जगह में) मेढक पत्थर रहेगा उसके नीचे जल मिलेगा।

**English:** In deserts, if the cane is visible then towards west of that tree nearly about 2.70 metre below ground level ground water veins flows which contains water.

Beneath the earth at about 1 metre below ground level if we see white frog later on yellow soil underlain by friable sandstone (about 2 m in thickness) then there is a ground water below it.

**Scientific Interpretation:**

Cane plants are hydrophytes, they can grow only on that part where more quantum of water will be available. Hence if in desertic area if we get cane tree which proves ground water worthy.

Frog needs moisture during hibernation state. Yellow soil and friable sandstone act as a porous zone for ground water percolation which proves the ground water worthy zone.

जम्ब्वाश्चोदग्घस्तैस्त्रिभिः शिराधो नरद्वये पूर्वा ।
मुल्लोहगन्धिका पाण्डुरा च पुरुषेऽत्र मण्डूकः ॥ 8 ॥

**Cane Tree**

**हिंदी:** जम्बू (जामुन वृक्ष) के उत्तर तीन हाथ पर दो पुरुष के नीचे शिरा है। पुरुषभर (1.80 metre nearly 12 feet or 3.60 m. below ground level) जमीन के नीचे खोदने पर लोहे के समान गंध वाली मिट्टी व उसमे मेधा चिन्ह दिखाई पड़ेगा।

**English:** Towards the North of Jamun tree and about 1.5 m. away there is a ground water flowing vein in the east direction. Excavating about 1.80 m below ground level the soil with the ferruginous smell with the presence of frog is available.

## Scientific Interpretation:

In Jamun fruits iron content is very high which indicates the nature of soil that is ferruginous soil. This soil is Porus and the rate of infiltration will be high at these places porous the ground water worthy zone.

जम्बूबृक्षस्य प्राग्वल्मीको यदि भवेत् समीपस्थः ।
तस्माद्दक्षिणपाश्वें सलिलं पुरुषद्वये स्वादु ।। 9 ।।

**हिंदी:** जामुन वृक्ष के पूर्व में समीप ही वल्मीक (सर्प स्थान) हो तो उसके दक्षिण दो पुरुष नीचे (12 feet or 3.60 metres) – जमीन के अन्दर मीठा जल मिलता है।

**English:** Towards the east from Jamun tree if we get ants head or snakes head then towards south from this point about 12 feet or 3.60 m below ground level proves the presence of fresh quality of water.

अर्धपुरुषे च मत्स्यः पारावतसन्निभश्च पाषाणः ।
मृःद्वति चात्र नीला दीर्घं कालं च बहु तोयम् ।। 10 ।।

**Jamun Tree**

**हिंदी:** जिस स्थान पर आधा पुरुष (3 feet or 1 meter) खोदने पर मछली, सफेद गोल पत्थर एव नीली मिट्टी मिलती है वहाँ अथाह जल होता है तथा वह बहुत समय तक चलता है।

**English:** After excavating about a metre fish, white bouldery stone and blue soil if exist then at that place good quantum of water with good quality is available for longer period.

## Scientific Interpretation:

Bouldery formations indicates about alluvial deposits, fish and blue soil is an indication of moisture zone under such situation sub surface aquifer will be productive zone with good quality.

पश्चादुदुम्बरस्य त्रिभिरेव करैर्नरद्वये सार्धं ।
पुरुषे सितोऽहिरश्माञ्जनोपमोऽधः शिरा सुजला ॥ 11 ॥

**हिंदी:** जल रहित देश में यदि गुलर का वृक्ष हो तो उसके तीन हाथ पश्चिम (4.5 feet or 1.35 m) अढाई पुरुष के नीचे (15 feet or 4.50 m) सुंदर जलवाहक शिरा मिलती है। एक पुरुष (6 feet or 1.80 m) खोदने पर सफेद सर्प दिखाई पड़ेगा फिर काजल के समान काला दिखाई पड़ने के बाद उसके नीचे जल जाने ।

**English:** In the desertic area, if we get Gular tree than about 1.35 m towards west and 4.50 m. below ground level ground water flowing vein occur. After digging 1.8 m. White snake followed by blackish soil occur then there is possibility to get ground water.

## Scientific Interpretation:

Gular tree indesertic area only will grow at that part where ground water will be available. White snakes proves moisture zone then blackish soil will be the another supporting evidence to get ground water.

उदगर्जुनस्य दृश्यो वल्मीको यदि ततोऽर्जुनाद्धस्तैः ।
त्रिभिरम्बु भवति पुरुषैस्त्रिभिरर्धसमन्वितैः पश्चात् ॥ 12 ॥

श्वेता गोधार्धनरे पुरुषे मृद्ङ्सरा ततः कृष्णा ।
पीता सिता ससिकता ततो जलं निर्दिशेदमितम् ॥ 13 ॥

**Arjun Tree**

**हिंदी:** बोखारे के वृक्ष के नीचे वल्मीक दिखाई पड़े तो वहाँ सवा तीन पुरुष (20 feet or 6 meters) जमीन के नीचे खोदने से पश्चिम तरफ जल शिरा बहती है।

भूजल प्राप्ति की पहचान यह है की आधा पुरुष (3 feet or about a metre) खोदने पर सफेद गोधा (white snail) एवं करीब (6 feet or 2 metre) खोदने पर सफेद काली मिट्टी तत्पश्चात भू जल प्रचुर मात्रा में उपलब्ध है।

**English:** Under the tree of Bokhara if ant's head is visible then at the depth of 6 metres towards west ground water vein flows.

To get a ground water the sub surface features viz about a metre below ground level availability of white snail then about 2 metre below ground level availability of white black soil then possibility of ground water exist.

**Scientific Interpretation:**

Bokhara tree grows in a wet zone which proves the soil quality and good aquifer zone beneath it.

To get a mollusea is a significant sign which proves the quality and quantity both the ground water.

वल्मीकोपचितायां निर्गुणचां दक्षिणेन कथितकरैः ।
पुरुषद्वये सपादे स्वादु जलं भवति चाशोष्यम् ॥ 14 ॥
रोहितमत्स्योऽर्धनरे मृत् कपिला पाण्डुरा ततः परतः ।
सिकता सशर्ककाऽथ क्रमेण परतो भवत्यम्भः ॥ 15 ॥

**हिंदी:** यदि मेवड़ी वृक्ष (कैथ) बाल्मीकी से युक्त हो तो उसके 3 हाँथ पर (4 feet or 1.20 metres) सवा दो पुरुष (13.5 feet or 4.5 m. below ground level) खोदने पर कभी न सूखने वाला मीठा जल पाया जाता है।

अर्धपुरुष (3 feet about a metre below ground level) के नीचे रोहित (रोहू मछली, पीली मिट्टी उसके नीचे सफेद मिट्टी उसके नीचे पत्थर युत रेती, उसके नीचे जल होता है।

**English:** If we get Kaith (Mewari tree) encircled by ant's head then about 1.20 m away on surface and after excavating about 4.05 m. b.g.l. good quality and quantity of ground water is available.

About a metre below ground level we will get a fish which indicates aquatic medium.

Yellow soil, white soil underlain by friable sand then ground water will be available.

**Scientific Interpretation:**

Ants head is a signal of water bearing zones hence under these circumstances there are more chances to get ground water. If Kaith tree is also grown in and around this proves the ground water worthy zone.

Fish is a symbol of aquatic medium then under it yellow soil which is underlain by white soil and friable sand. These formations are excellent information to get the ground water.

पूर्वेण यदि वदर्या वल्मीको दृश्यते जलं पश्चात् ।
पुरुषैस्त्रिभिरादेश्यं श्वेता गृहगोधिकार्द्धनरे ।। 16 ।।

**हिंदी:** यदि वेर के पेड़ के पूर्व बल्मीक हो तो उसके पश्चिम 3 हाथ (4.5 feet or 1.35 m) पर जल मिलेगा। वहाँ की निशानी यह है कि आधे पुरुष (3 feet or 0.90 m) पर सफेद छिपकली (White Lizard) मिलेगी तथा उसके नीचे खोदने पर जल मिलेगा।

**English:** If towards east of the ber tree ant's head exist then towards west ant the distance of 1.35 m we will get water. The symbol is that we will get white lizard in the depth of about 1 metre beyond this point possibility exists to get ground water.

**Scientific Interpretation:**

As ber is a thorny plant to get ants head is a symbol of moisture and the reptiles group lives in moisture. These indications proves the availability of good water.

सपलाशा बदरी चेद् दिश्यपरस्यां ततो जलं भवति ।
पुरुषत्रये सपादे पुरुषेऽत्र च दुण्डुभश्चिह्नम् ।। 17 ।।

**हिंदी:** यदि पलास और बेर वल्मीक युक्त हो या न हो तो बेर वृक्ष के 3 हाथ (4.5 feet or 1.35 m) पश्चिम सवा तीन पुरुष के नीचे (20 feet or 6 m. b.g.l.) खोदने पर जल मिलेगा । जिसका चिन्ह यह है कि (6 feet, 2 m. b.g.l.) बिना विषवाला दो मुख का सर्प मिलेगा, उसके पास ही जल मिलेगा।

**English:** If palash and Ber associated with ant's head or not even then 1.35 m. away towards west from Ber tree after excavation at a depth of 6 m. b.g.l.

We will get water. Another symbol is that at a depth of 2 m. b.g.l. a bifacial non poisonous snake will be seen adjacent to this strata proves the presence of water.

**Scientific Interpretation:**

Palash and Ber are aquatic plants and bifacial snake is an aquatic animal. Presence of ant's head and deep moisture proves the presence of ground water.

विल्वोदुम्बरयोगे विहाय हस्तत्रयं तु याभ्येन ।
पुरुषैस्त्रिभिरम्बु भवेत् कृष्णोऽर्द्धनरे च मण्डूकः ॥ 18 ॥

**Bel + Gular together**

**हिंदी:** जहाँ बेल गुलर दोनों का योग एक साथ हो उसके 3 हाथ (4.5 feet or 1.35 m) दक्षिण छोड़कर तीन पुरुष नीचे (18 feet or 5.40 metre) खोदने पर जल मिलेगा। अर्ध पुरुष (3 feet or 0.90 m.) खोदने पर काला मेद्या मिलेगा।

**English:** At a place where Bel and Gular both trees are available together, under these conditions about 1.35 m. away towards south at a depth of 5.40 m. chance of ground water. At this place at a depth of 0.90 m. b.g.l. black frog will be available.

**Scientific Interpretation:**

Bel and gular are aquatic in character which grow only in the wet and moist zone proves ground water worthy zone.

काकोदुम्बरिकायां वल्मीको दृश्यते शिरा तस्मिन् ।
पुरुषत्रये सपादे पश्चिमदिकस्था वहति सा च ॥ 19 ॥
आपाण्डुपीतिका मृद्गोरसवर्णश्च भवति पाषाणः ।
पुरुषार्धं वु मुदनिभो दृष्टिपथं मूषको याति ॥ 20 ॥

**हिंदी:** बोखारे के पेड़ के नीचे वाल्मीकी यदि दिखाई पड़े तो वहा सवा तीन पुरुष (20 feet or nearly 5 m. b.g.l.) जल शिरा बहती है । चिन्ह यहाँ सफेद मिट्टी उसके नीचे पीली उसके नीचे श्वेत पत्थर एवं आधापुरुष (3 feet or 1 metre) खोदने पर सफेद चूहा दिखाई पड़ता है।

**English:** If ant's head under the Bokhara plant then 20 feet or nearly 5 m. depth b.g.l. a ground water vein flows. On surface white soil, beneath it yellow soil after that white stone are visible. At a depth of 1 m. excavation chances exist to get the white rat.

**Scientific Interpretation:**

Bokhara is an aquatic plant and presence of ants head and soil (white, yellow) and friable sand stone is an indication to get ground water.

# CHAPTER 2

जलपरिहीने देशे वृक्षः कम्पिल्लको यदा दृश्यः ।
प्राच्यां हस्तत्रितये वहति शिरा दक्षिणा प्रथमम् ।। 21 ।।

**हिंदी:** जल रहित देश में कपिल (रुई, मदार बणी मौसरी) का पेड़ दिखाई पड़े तो इसके पूर्व तीन हाथ पर (4.5 feet or 1.35 m) हाथ के नीचे दक्षिण शिरा बहती है।

**English:** In the desertic area if cotton, mandar and mausari trees are visible then towards east about 1.35 m away and at a depth of 1.50 m. b.g.l. ground water carrying vein form south flows.

**Scientific Interpretation:**

In the desertic area if the tree will grow then this it self is a symbol of moisture zone and porous formation proves the chances to get the ground water.

मृन्नीलोत्पलवर्णा कापोता दृश्यते ततस्तस्मिन् ।
हस्तेऽजगन्धको मत्स्यकः पयोऽल्पं सक्षारम् ।। 22 ।।

**हिंदी:** उपरी सतह पर नील कमल के समान मिट्टी फिर कबूतर के रंग की मिट्टी या पिली मिट्टी दिखाई पड़ेगी उसके नीचे एक हाथ खोदाई करने पर बदबुदार मछली दिखाई पड़ेगी।

**English:** Top of the soil layer is blue in appearance then followed by brown or yellow soil then after excavation nearly 0.5 m. b.g.l. a fish with bad smell will be visible.

## Scientific Interpretation:

Yellow and brown soil are good soil which are porous and rate of percolation is also high. It proves the presence of ground water.

शोणाकतरोरपरोत्तरे शिरा द्वौ करावतिक्रम्य ।
कुमुदा नाम शिरा सा पुरुषत्रयवाहिनी भवति ।। 23 ।।

**हिंदी:** शोणाक (टेसू) के वृक्ष के वायु कोण में दो हाथ छोड़कर तीन पुरुष के नीचे (3 feet or 1 m and 13 feet or 6 m. b.g.l.) जल वाहिनी शिरा बहती है।

**English:** In the N.E. direction from Tesu tree about a metre away a ground water bearing veins flows which is known as Kumuda.

## Scientific Interpretation:

Tensu is a plant of aquatic environment proves the ground water worthy zone.

आसन्नो वल्मीकी दक्षिणपाशर्वे विभीतकस्य यदि ।
अध्यर्धं भवति शिरा पुरुषे ज्ञेया दिशि प्राच्याम् ।। 24 ।।

**हिंदी:** विभीतक (वहेडा) वृक्ष के दक्षिण दिशा में वल्मीक देख पड़े तो उस वृक्ष के दो हाथ पूर्व में डेढ़ पुरुष के नीचे (9 feet or 2.70 m) जल शिरा जाने।

**English:** Towards the south of Bahera tree if ant's head is visible then about a metre any towards east at a depth of 2.70 m. b.g.l. ground water vein exist.

## Scientific Interpretation:

Bahera and ants head both are associated with ground water hence proves the positive indication of ground water.

तस्यैव पश्चिमायां दिशि वल्मिको यदा भवेद्वस्ते।
तत्रो दग भवति शिरा चतुर्भिरर्धाधिकैः पुरुषैः ।। 25 ।।

**हिंदी:** बहेणा के पश्चिम में बल्मीक हो तो उस वृक्ष के उत्तर 1.5 फीट (0.45 m) छोड़कर 30 फीट (9 m) के नीचे जलशिरा बहती है।

**English:** Towards the west from Bahera tree towards North 0.45 m. away if ant's head exist then there are water bearing veins at a depth of 9 m.

**Scientific Interpretation:**

Bahera is an aquatic plant and ants head indicates the moisture zone and the depth may be more which indicate the ground water promising zone.

**श्लोकः**

श्वेतो विश्वम्मरकः प्रथमे पुरुषे, तु कुङ्कमाभोऽश्मा ।
अपरस्यां दिशि च शिरा, नश्यतिवर्षत्रयेऽतीते ।। 26 ।।

**हिंदी:** एक पुरुष प्रमाण खोदने पर (6 feet or 2 metres) सफेद विश्वम्भरक (प्राणी विशेष) दिखाई पड़ता है, बाद में लाल पत्थर उसके नीचे पश्चिम में जल शिरा है लेकिन ३ वर्ष में इसका नाश होगा।

**English:** After digging about 2 metres b.g.l. some animal like structure is visible after that the red stone (Laterite or Red sand stone) is incurred. After that a vein which carries ground water flows from west but it is not sustaining for more than 3 years.

**Scientific Interpretation:**

Sand Stone or Laterite both are porous which will allow rain water to infilterate at faster rate. Perched conditions are also observed here – which shows poor sustainability of ground water which may vanish soon. The period may be 3 years or less.

सकुशः सित ऐशान्यां वल्मीको यत्र कोवि दारस्य ।
मध्ये तयोर्नैरर्ध पञ्चमैस्तो यम क्षोम्यम् ।। 27 ।।

**हिंदी:** केविदार (Bachna Purperia linn (Leguminosae)) का वृक्ष जिस भूमि में दृष्टि गोचर हो यदि उस वृक्ष से ईशान कोण में कुशा से

युक्त सफ़ेद रंग की बाल्मिक हो तो उस वृक्ष और वल्मीक के मध्य में साढे पांच पुरुष के नीचे (33 feet or 10 m. b.g.l.) बहुत जल रहता है।

**English:** Kovider tree is located on which ground then towards NE (Isan angle) a white ant's head associated with Kusha is visble then in between these two about 33 feet (about 10 m. b.g.l.) a good quantum of ground water repository exist.

## Scientific Interpretation:

Kachnar is hydrophytes and white ant's head is an indicator of ground water bearing zones from deep sources. Hence these two symbols are an indication of ground water.

## श्लोकः

प्रथमे पुरुषे भुजगः कमलोदरः सन्निभो मही रक्ता ।
कुरु विन्दः पाषाण शिचह्वान्ये तानि वाच्यानि ॥ 28 ॥

**Ants Head under Kush Tree**

**हिंदी:** प्रथम में कमलोदर तुल्य सर्प दिखाई पड़ेगा उसके बाद (लाल) वर्ण की भूमि मिलेगी, फिर हरे रंग का पत्थर मिलेगा । ये सारे चिन्ह उस जगह के बारे में हैं।

**English:** At beginning we will see lotus colour snake (at bottom) towards stomach side then red color soil after that green stone will be visible.

**Scientific Interpretation:**

Red soil – Laterite which is porous, allows ground water to percolate and supports for red appearance from bottom side for snake.

More moistures in this soil which would have been allowed the bottom rock to convert in to green colour due to algal environment.

**श्लोकः**

यदि भवति सप्तवर्णो वल्मीक वृतस्त दुत्तरे तोयम् ।
वाच्यं पुरुषैः पञ्चमिरत्रापि भवन्ति चिन्हानि ॥ 29 ॥

**हिंदी:** यदि सप्तवर्ण वृक्ष वल्मीक से युक्त हो तो उस वृक्ष से उत्तर ५ हाँथ के बाद (10 feet or 3 m) जल होता है उसके चिन्ह इस प्रकार के होते है (अगले श्लोक में)

**English:** If the saptavarna tree associated with ant's head then towards North about 10 feet (3 m) away, chances exist for ground water.

**Scientific Interpretation:**

Saptavarna tree is a hydrophytes & ant's head is a symbolic guide line to predict the presence of ground water.

**श्लोकः**

पुरुषार्ध मण्डूको हरितो, हरितालसन्निभा भूश्च ।
पाषाणोऽभ्रनिकाशः सौम्या च शिराशुमाम्बुवहा ॥ 30 ॥

**हिंदी:** पतले ½ पुरुष के नीचे हरित वर्ण का मेढक मिलेगा, फिर हरिताल के समान भूमि मिलेगी, उसके बाद मेघ के तुल्यवर्ण का पत्थर मिलेगा, उसके नीचे मीठे जलवाली उत्तरी शिरा होगी।

**English:** About 3 feet (a metre) b.g.l. a frog of green appearance will be visible then the soil appearance will be like a green pond then a cloudy colour stone will be visible, Then beyond that there will be good quality of water flowing vein which comes from North.

## Scientific Interpretation:

The green frog is a symbol of soil moisture zone where % of soil moisture is very high. The cloudy stone can be correlated with sand stone which is having the capacity of good ground water repositories if fractured. Scientifically this have been observed in the field area.

## श्लोकः

सर्वेषांवृक्षाणामधः स्थितो दर्दुरो यदा दृश्यः ।
तस्मादस्ते तोयं चतुर्भिरर्धाधिकैः पुरुषैः ॥ 31 ॥

**हिंदी:** सब वृक्षों में जिनके नीचे मेढक हो, उस वृक्ष के उत्तर दिशा में एक हाथ दूरी पर (3 1/2 feet या 1 मीटर) दूरी पर 4 1/2 पुरुष (27 feet or करीब 9 metre) जमीन के नीचे जल रहता है।

**English:** Among all the trees under which the frog sits generally from that tree about 1 m. towards North & 9 m. b.g.l. the ground water occurs.

## Scientific Interpretation:

Frog are cold blooded animals hence they are only visible at those places where soil moisture content is high hence chances occurs that at lower depth the ground water will occur.

## श्लोकः

पुरुषे तु भवति नकुलो, निला मृत्पीतिका ततः श्वेता ।
दर्दुर समान रुपः पाषाणो, दृश्यते चात्रः ॥ 32 ॥

**Frog under Peepal Tree**

**हिंदी:** एक पुरुष नीचे (6 फीट or 1.80 m. b.g.l.) नेवला उसके बाद नीली मिट्टी उसके बाद स्वेत मिट्टी उसके बाद मेढ़क के समान पत्थर मिलेगा उसके नीचे जल मिलेगा।

**English:** About 1.80 m. b.g.l, Mongoose will be visible, after that blue soil then white soil, below that the stone just like as a frog colour appearance will be visible under that the ground water occurs.

**Scientific Interpretation:**

Mongoose is visible under soil moisture zone thru blue or white soil which follows granite or ferruginous sand stone? These rocks are good for ground water as more cracks and fissures are developed which proves the presence of ground water.

यद्यहिनिलयो दृश्यो दक्षिणतः, सांस्थितः करञ्जस्य ।
हस्त द्वये तु याम्ये, पुरुष त्रितये शिरा सार्ध ॥ 33 ॥

**हिंदी:** करंज वृक्ष के दक्षिण में (दिठोरी - Pongama glaberavenrt (Leguminosae)) सर्पावास का वल्मीक दिखाई दे तो उस करञ्ज वृक्ष

के दक्षिण दिशा में दो हाथ दूरी के बाद 3 1/2 पुरुष नीचे (21 feet, 6.30 metres) जलवाहिनी सोती मिलती है ।

**English:** Towards the south of Karanj tree if the snake head is invisible then 1.5 m away towards south at a depth of 6.30 m. b.g.l. a ground water flowing vein (aquifer) is available.

**Scientific Interpretation:**

The Karanj tree grow in moisture zone and snake head is a symbol of moisture. At these places soil porosity will be very high based on the symptoms there are chances to get ground water in aquifer-zone.

**श्लोकः**

कच्छपकः पुरुषार्द्धं प्रथमं चोद्भिद्यते शिरा पूर्वा ।
उदगन्या स्वादुजला हरितोऽश्माद्यस्ततस्तोयम् ॥ 34 ॥

**हिंदी:** प्रथम आधे पुरुष नीचे खोदने पर कच्छप उसके आगे पूर्वाभिमुख जल बहता हुआ मिलेगा द्वितीय उत्तर शिरा मीठा जल मिलेगा, उसके नीचे – हरे रंग का पत्थर मिल करके अन्त में जल मिलेगा।

**English:** (in continuation)

About one metre after digging tortoise then a water flowing vein from east will be visible, then from north good quality of water will appear then green stone underlain by ground water will be visible.

**Scientific Interpretation:**

Tortoise is aquatic animal green stone (by weathering act on sand stone have been converted into green colour. Based on the geological concept sand stone is a good ground water repository.

**श्लोकः**

उत्तरतश्च मधूका दहि निलयः, पश्चिमे तरोस्तोयम् ।
परि हृत्यपञ्च हस्तान्, धाष्टमपौरुषान प्रथमम् ॥ 35 ॥

**हिंदी:** महुआ (Bassia Latifalia Roxli (Saptaceae)) के वृक्ष के उत्तर दिशा में यदि सर्प का वल्मीक दिखाई दे तो उस वृक्ष से पश्चिम और पाँच हाथ की दूरी के बाद साढ़े सात पुरुष नीचे (45 फीट, 13.51 m) जल रहता है।

**English:** Towards the North from the Mahua tree if snakes head is visible then towards the west about 3 m away and 13.5 m. below ground level – water is available.

**Scientific Interpretation:**

Mahua tree grows in soil moisture zone – where % of moisture content is high snakes head will be visible only at those places where moisture content will be high. This is the indication of aquatic environment & more chances exist for ground water repository.

**श्लोकः**

अहिराजः पुरुषेऽस्मिन्धूमा, धात्री कुलत्थवर्णोऽश्मा ।
माहेन्द्री भवति शिरावहति, सफेनं सदातोयम् ॥ 36 ॥

**हिंदी:** प्रथम पुरुष के नीचे सर्पराज मिलेंगे तदन्तर उसके नीचे धूम्रवर्ण की भूमि तथा उसके नीचे कुलथ वर्ण का पत्थर मिलेगा। वहाँ पर माहेन्द्री नाम पूर्वाभिमुख जल बहने वाली शिरा होती है तथा उससे सदैव फेन युक्त जल बहा करता है।

**English:** About 2 m below ground level snake will be visible, beyond that smoky colour soil the brownish black colour stone will be visible. At this depth we will get a ground water flowing vein which is known as mahendri by which always water flows.

**Scientific Interpretation:**

Smoky colour soil, snake visibility and brown colour stone shows the symbol of water logged area. Here at depth good aquifers exist by which we may get perrinial water with good yield.

श्लोकः

वल्मीकः स्निग्धो दक्षिणेन, तिलकस्य सकुश दूर्वश्चेत ।
पुरुषैः पञ्चमिरभ्भा दिशि, लारुण्यां शिरा पूर्वा ।। 37 ।।

**हिंदी:** जो तिलक वृक्ष (Rubiacae family) से दक्षिण में उत्तम वल्मीक जहाँ पर कुश-दूब से युक्त हों, तो उस वृक्ष से पश्चिम और पाँच हाथ के बाद (8 feet or 2 m) पाँच पुरुष (30 feet, nearly 9 m) पर जल रहता है तथा वहाँ पूर्वा शिरा बहती है।

**English:** Towards the south form the Tilak tree if good ant's head is visible where Kush and Durva (grass) also exist then towards the west from that tree about 2 m. away at a depth of 9 m. b.g.l. ground water exist. This is known as eastern flowing vein.

**Scientific Interpretation:**

Ant's head Kush or Durva will grow why in porous formation & where soil moisture exist. Such formations proves that rate of infiltration will be high in these areas & good ground water zones will exist in this formation.

श्लोकः

सर्पावासः पश्चाद्यदा, कदम्बस्य दक्षिणेन जलम् ।
परतो हस्तत्रितयात् षडभिः पुरुषैस्त री योनैः ।। 38 ।।
कोवेरी चात्र शिरा वहति, जलं लोह गन्धि चाक्षोभ्यम् ।
कनक निभो मंडूको, नरमात्रे मृत्रिका पीता ।। 39 ।।

**हिंदी:** यदि सर्पावास वल्मीक कदम्ब वृक्ष (Anthocephalus cadamba Murr (Rubiaceae)) से पश्चिम दिशा में हो तो जल दक्षिण दिशा में तीन हाथ (4.5 feet or nearly 1.5 m), पौने छः पुरुष (35 feet or nearly 10 m) के नीचे बहता है। वहाँ उतरा नामक शिरा होती है, उसमें लोहे के समान गंध निकलता हुआ जल निकलता है। उसका चिन्ह वही है कि प्रथम एक पुरुष के नीचे खोदने से (6 feet or nearly 2 m) सोना के समान मेढक निकलता है और पीतवर्ण की मिट्टी निकलती है।

**Scientific Interpretation:**

Towards the west from the Kadamb tree, if the snakes head is visible then the ground water is available about 10 m. b.g.l. towards south about 2 m ahead. The vein which flows is known as uttara and iron smell appears there.

At this depth we get a frog of golden appearance at the depth of 2 m. b.g.l. which follows with yellow soil.

**Scientific Interpretation:**

Frog/snakes head is the symbol of porous formation where soil moisture exist.

Yellow soil is the symbol of recent deposits or younger alluvial formation where good ground water aquifer will exist.

Above all symptoms indicate good ground water zones availability in such type of formations.

### संस्कृतः

बल्मीक संवृतो यदि तालोवा, भवति नालिके रोवा ।
पश्चात् षsभिर्हस्तैर्न रेश्चचतुर्भिः शिरा याम्या ॥ 40 ॥

**हिंदी:** यदि ताड़ (Borossus flabellifer - Linn (Palmea)) के वृक्ष या नारियल (cocconut) के वृक्ष के समीप वल्मीक स्थित हो तो उन वृक्षों के पश्चिम दिशा में 6 हाथ (9 feet or 3 meter) के बाद चार पुरुष (24 feet or 8 m. b.g.l.) के नीचे जल दिखाई पड़ेगा और वहाँ दक्षिण शिरा होगी।

**English:** Adjacent to Palm or coconut tree if ant's head is visible then towards the west from that tree about 3 metre ahead at a depth of 8 m. b.g.l. ground water will be available and the ground water will flow from southern vein.

**Scientific Interpretation:**

The palm and Coconut tree will grow only, soil moisture zone & the ant's head will be visible only in porous formation, indicates ground water promising zone in this area.

**Snakes hole under Coconut Tree**

# CHAPTER 3

**श्लोकः**

याम्येन कमिस्थं स्याहिसं, ध्रयश्चे दुदग्जलं वाच्यम् ।
सप्त परित्यज्य करान खात्वा, पुरुषान जलं पञ्च ।। 41 ।।

**हिंदी:** यदि सर्पावास वल्मीक कैथ के वृक्ष से दक्षिण दिशा में हो तो उस कैथ वृक्ष से उत्तर दिशा में 7 हाथ (10.5 feet, 3.5 m) के बाद 5 पुरुष (9 m) नीचे जल रहता है।

**English:** If the snakes head is towards south from the Kaith tree then absent 3.5 on north from that tree and about 9 m. below ground level chances exist to get ground water.

**Scientific Interpretation:**

The snakes head shows the quality of soil which indicates more porous formation. If the porosity is high then rate of infiltration will also be high which shows availability of ground water.

Kaith tree is also aquatic plant which shows the presence of water in the area. Ground water worthy zone exist here.

**श्लोकः**

कर्बुरकोऽहिः पुरुषे कृष्णामृत् पुटभिदापि च पाषाणः ।
श्वेतामृत पश्चिमतः शिरा ततश्चोत्तरा भवति ।। 42 ।।

**हिंदी:** उसके ये चिन्ह हैं

प्रथम तो एक ही पुरुष (6 feet, 1.80 m) कर्बुर वर्ण (कबूतर के रंग का) सर्प तथा काली मिट्टी होती तदन्तर पुट भेदी पत्थर मिलेगा तथा

आगे श्वेत मिट्टी तथा वहाँ ही एक पश्चिम शिरा तथा उसके बाद उतारा शिरा होगी।

**English:** About 1.80 m. b.g.l. a snake of Pigeon Colour associated with black soil which followed by friable stone and white soil. At the same point ground water vein flowing from west and then north vein will be available.

## Scientific Interpretation:

Availability of snake shows the porous formation where rate of infiltration will be high. At the same time friable stone (shows fractures) and white soil proves the ground water appearance under such conditions.

## श्लोकः

अश्मन्तकस्य वामे बदरीवा, दृश्यतेऽहिनिलयोषा ।
षड्भिरुदक तस्य करैः स्वार्ध पुरुषत्रये तोयम् ॥ 43 ॥

**हिंदी:** जो बदरी वृक्ष या सर्प गृह अश्मन्तक वृक्ष के उत्तर या बाये और हो तो जल अश्मन्तक वृक्ष से 6 हाथ (9 feet or nearly 3 m) के बाद साढ़े तीन पुरुष (21 feet or 7 m. b.g.l.) उत्तर दिशा में होता है।

**English:** Badri tree which is located near the snakes head which may be either north or left then ground water will be available about 9 feet or nearly 3 m and 21 feet or nearly 7 m. b.g.l. towards north the ground water flowing vein will be available.

## Scientific Interpretation:

Snakes head shows the porous soil & Badri tree from aquatic nature and this formation will be suitable for ground water occurrence.

## श्लोतः

कूर्मः प्रथमें पुरुषे, पाषाणो धूसरः ससिकता मृत् ।
आदौ च शिरा याम्या, र्श्वोत्तरतो द्वितीया च ॥ 44 ॥

**हिंदी:** और इस चिन्ह से उसको ज्ञात करें प्रथम एक पुरुष (6 feet or 1.8 m) खोदने पर कच्छप बाद में धूसर रंग का पत्थर होगा और अंत में बालुका युक्त मिट्टी मिलकर पहले दक्षिण शिरा, अंत में ईशान कोण में बहने वाली दूसरी शिरा मिलेगी।

**English:** After excavating top – at a depth of 1.8 m a tortoise will be visible then soil colour stone and at least sandy soil will be available. A south flowing vein and another vein in Ishan angle (North East) will also be available.

## Scientific Interpretation:

Tortoise is an aquatic animal and sandy soil is a significant soil from aquifer point of view. The above symbols mile prove the great water bearing zone in these formations.

## श्लोकः

वामेन हरिद्रतरोर्वल्मीक, श्चेज्जलं भवति पूर्वे ।

हस्तत्रितये संत्रयंशैः पुम्भिः पञ्चभिर्भवति ॥ 45 ॥

**हिंदी:** यदि हल्दी के वृक्ष से उत्तर और बल्मीक देख पड़े तो उस वृक्ष से पूर्व दिशा में जल 3 हाथ (4.5 feet or 1.5 m) के बाद साढ़े पाँच पुरुष के नीचे (33 feet or nearly 10 m. b.g.l.) रहता है।

**English:** If towards the north from Haldi tree (Temarind) if ant's head is visible then towards east about 1.5 m ahead at the depth of 33 feet or 10 m. b.g.l. ground water will be available.

## Scientific Interpretation:

Haldi tree is a hydrophytes & if ant's head are also visible which indicates high porosity of the soil.

The above situations proves ground water worthy zone.

## श्लोकः

नीलो भुजगः पुरुषे, मृत्पीता मरकतो पमश्चाश्मा ।

कृष्णा भूः प्रथमं वारुनी, शिरा दक्षिणे नान्या ॥ 46 ॥

**हिंदी:** प्रथम वहाँ पर एक पुरुष के नीचे जल दिखाई पड़ेगा (6 feet or 2 m.) उसके बाद पीली मिट्टी तदन्तर मरकत पत्थर मिलेगा उसके बाद काली भूमि उसके बाद प्रथम पश्चिम शिरा तथा दूसरी दक्षिण शिरा होगी।

**English:** At first at the depth of about 2 m. below ground level, yellow soil then sand stone will be observed. After that black soil then ground water flowing veins west and south will be seen there.

### Scientific Interpretation:

Yellow soil and sand stones are ground water bearing horizons, hence definitely these symbols are the supporting evidences to get the ground water.

### श्लोक:

जल परिहीने देशो दृश्यन्ते, ऽनूपजानि चोन्नि मित्तानि ।
वीरण दूर्वा मृदवश्च, यत्र तस्मिन जलं पुरुषे ।। 47 ।।

**हिंदी:** जल परिहीन देश में जहाँ पर अनूप चिन्ह दिखाई पड़े वहाँ अधिक जल कहना चाहिये, प्रथम उसके उपर वीरण दूर्वा हो तो वहाँ की भूमि अति कोमल होती है एवं जल एक पुरुष (6 feet or 2 m) के नीचे होता है।

**English:** In a desertic land where Annop symbol is visible then there are chances to exist more ground water. If the grass is available on the top then the soil is very soft and water is available at the depth of 2 m. b.gl.

### Scientific Interpretation:

Availability of grass in a desertic soil indicates the high moisture %. At the same time where soil is also friable indicates it is more porous indicates chances are better to get ground water.

### श्लोक:

भाङ्गी त्रिवृता दन्ती, सूकरपादी च लक्ष्मणा चैव ।
नवमाक्तिका च हस्तद्वये, ऽम्बु याम्ये त्रिभिः पुरुषैः ।। 48 ।।

**हिंदी:** धरहम दंडी, नीशोध (विधारा), वज़्रदत्री, केवाच, लक्ष्मणा, निवारी के वृक्ष हो तो उसके दक्षिण और जल दो हाथ के बाद (3 feet or 1 m), 3 पुरुष (13 feet, 6 m. b.g.l.) नीचे रहता है।

**English:** If the above mentioned trees are visible then towards south about 1 m away at the depth of 6 m. b.g.l. ground water will be available.

**Scientific Interpretation:**

These trees are ground water indicator and grow only at those places where water is available & proves the availability of ground water.

**श्लोकः**

स्निग्धाः प्रलम्बशाखावामन विकटद्रुमाः समीपजलाः ।
सुषिरा जर्जर पत्रा रुक्षाश्च, जलेन सन्त्यकताः ।। 49 ।।

**हिंदी:** जो वृक्ष चिकना, विशाल शाखाओं से युक्त, अतिसय छोटा तथा विस्तीर्ण हो उसके समीप जल रहता है। जो वृक्ष सारहीन, जर्जर पत्र तथा रूखे हैं, उनके पास जल नहीं रहता।

**English:** A tree which is smooth, branched and stout (with less height) water is available adjacent to such type of tree. Tree which is rough and without leaves, water would not be available adjacent to such tree.

**Scientific Interpretation:**

A tree with numerous branches and leaves is an indication of more ground water availability to the roots of such trees. Hence under such conditions definitely ground water will be available.

And adjacent to dry tree (no leaves in the tree) shows the dry zone where no chance of ground water.

**श्लोकः**

तिलकाम्रातकवरुणकभल्लातक, विल्वतिन्दुकाङ्कोलाः ।
पिण्डारशिरीषाञ्जनपरुषका, वञ्जुलोऽतिबला ।। 50 ।।

**हिंदी:** यदि तिलक, आम्रातक, मल्लातक, बिल्ब, तिन्दुक, अङ्कोल्क, पिंडार, शिरीष, अन्जन वृक्ष, परूषक वृक्ष, बज्जुक, अतिबला होता...

**English:** The above said trees. Tilak, Amratak, Mallatak, Bilb, Tinduk, Ankol, Pindar, Shirish, Arjun tree, Parushak tree, Barful, Atibela.

## Scientific Interpretation:

The above trees are hydrophytes & they grow only in such condition where ground water condition are quite good.

## श्लोकः

एते यदि सुस्निग्धा वल्मीकैः परिवृता स्ततस्तोयम् ।
हस्तेस्त्रिभिरुत्तरतश्च, तुर्भिरर्धेन च नरेण ।। 51 ।।

**हिंदी:** यदि ये वृक्ष वल्मीक से युक्त हो अथवा वल्मीक इनके समीप हो तो इन वृक्षों के उत्तर दिशा में ३ हाथ (4.5 feet or 1.5 m) के बाद साढ़े चार पुरुष (27 feet or 9 metre) जमीन के नीचे जल रहता है।

**English:** If these tree are covered by ant's head or these are located nearly by then towards the north about 1.5 m ahead at the depth of 9 m. b.g.l. ground water exist.

## Scientific Interpretation:

Ant's head and hydrophytes tree are symbolic for good ground water bearing zones known as aquifers.

## श्लोकः

अतृणे सतृणा यस्मिन सतृणे, तृणवर्जिता मही यत्र ।
तस्मिन् शिरा प्रदिष्टा, वक्तव्यं वा धनं चास्मिन ।। 52 ।।

**हिंदी:** तृण रहित भूमिपर जहाँ तृण उपजा हो अथवा तृण संयुक्त भूमि पर जहाँ तृण न हो उस स्थान पर जल तथा धन रहता है।

**English:** At a place where there is no grass, the grass is grown there and on the grassy land where there is no grass is an indication of water and wealth.

## Scientific Interpretation:

In a desert where generally grasses will not grow – there if grasses grows indicates possibility of good water sources.

Vice – Versa if reverse situation occur indicates water logging condition due to that grass will not occur.

## श्लोकः

कण्टक्यकण्टकानां व्यत्यासेऽम्भस्त्रिभिः करै पश्चात् ।
खात्वा पुरुषत्रितयं, त्रिभात्रयुक्तं धनं वा स्यात् ॥ 53 ॥

**हिंदी:** जहाँ पर कंटक वृक्ष में अंकटक वृक्ष हो अथवा अंकटक वृक्ष में कंटक वृक्ष हो उस वृक्ष के पश्चिम दिशा में ३ हाथ दे बाद (5 feet - about 2 metres), ३¼ पुरुष के नीचे (20 feet about 6 metres) पर जल मिलता है।

**English:** In the area of thorny trees where trees are grown without thorn or vice – versa means trees without thorn are there at that place – thorny trees are grown then in the west direction of this tree after 2 metres at the depth of 6 m. b.g.l. water is available.

## Scientific Interpretation:

Thorn is a cause of desert means

   i. if at that place without thorn (bushy) plant will grow give on indication of water.
   ii. In bushy plant area (without thorn plant) if – plant with thorn will grow give an indication of ground water.

## श्लोकः

नदति मही गम्भीरं यस्मिंश्चरणाहता जलं तस्मिन् ।
सार्धैस्त्रिभिर्मनुष्यैः कौबेरी तत्र च शिरा स्यात ॥ 54 ॥

**हिंदी:** जिस स्थान पर भूमि को चरण से मारने पर गम्भीर शब्द हो, वहाँ उत्तर शिरा बहती है और ३½ पुरुष (20 feet - about 6 meters b.g.l.) खोदने पर जल रहता है।

**English:** At a place where the earth will be depressed and the sound will appear – that is heavy sound, indicates north veins flows at that point & the water will be available at the depth of 6 metres b.g.l.

**Scientific Interpretation:**

The sound of depression that is heavy is a symbol of moisture content in the soil. If the moisture % will be high, the possibility of ground water will also be very high.

श्लोक:

वृक्षस्यै का शाखा यदि विनता, भवति पाण्डुरा वा स्यात् ।
विज्ञातव्यं शाखातले, जलं त्रिपुरुषं खात्वा ॥ 55 ॥

**हिंदी:** यदि वृक्ष की एक शाखा झुकी हुयी हो या पाण्डूर वर्ण की हो तो उस डाल के नीचे तीन पुरुष नीचे (18 feet or 6 metre b.g.l.) पर जल मिलेगा।

**English:** If the one branch of the tree is inclined or its colour is dark red then there are chances to get a ground water at a depth of 6 m below ground level.

**Scientific Interpretation:**

Inclination of the tree may be due to gravitational forces. This force direction is due to ground water flow direction indicates the presence of ground water.

श्लोकः

फक कुसुम विकारोयस्य, तस्य पूर्व शिरा त्रिभिर्हस्तैः ।
भवति पुरुषैश्चतुर्भिः, पाषाणोऽधः शितिः पीता ॥ 56 ॥

**हिंदी:** जिस वृक्ष के नीचे फल या पुष्प का अन्य किसी वृक्ष का विकार हो तो उस वृक्ष के पूर्व में तीन हाथ आगे (4.5 feet, 1.5 metres) ४ पुरुष (24 feet - nearly 8 m. b.g.l.) खोदने पर जल है।

**English:** Under the tree where another tree having fruits/flowers are seen, then in east direction from that tree about 1.5 m away at a depth of 8 metres b.g.l. ground water is available. After digging at first stone will be observed then yellow soil will occur.

**Scientific Interpretation:**

A tree under tree having fruits/flowers – shows the indication of productive soil with good ground water bearing zone.

श्लोकः

यदि कण्टकारिका कण्डकैर्विता, दृश्यते सितैः कुसुमैः
तस्यास्तलेऽम्बु वाच्यं, त्रिभिर्तंरैरर्धपुरुषे च ।। 57 ।।

**हिंदी:** यदि कंटक वृक्ष कांटे से रहित हो या सफ़ेद फूल दिखाई पड़े तो उसके नीचे 3½ पुरुष (21 feet or 7 meter) पर जल मिलता है।

**English:** If thorny trees are without thorn or white flowers are visible then at this place at a depth of 7 metre b.g.l. water will be available.

**Scientific Interpretation:**

Thorny tree without thorn is an indication of present of ground water in soil or proves soil is porous and rate of infiltration will be high. It proves that good productive zone will be visible at lower depth.

श्लोकः

खर्जूरी द्विशिरस्का यत्र, भवेज्जल दिवर्जिते देश ।
तस्याः पश्चेमभागे, निर्देश्यं त्रिपुरषैर्वारि ।। 58 ।।

**Dates with two heads**

**हिंदी:** जिस निर्जल देश में दो शिर का (दो डाल का) खजूर का वृक्ष हों, उस खजूर वृक्ष के पश्चिम में २ हाथ के बाद (4 feet or 1.5 m) तथा ३ पुरुष के नीचे (18 feet or 6 metres b.g.l.) जल मिलता है।

**English:** In desertic area – a Date tree with two branches if visible then towards west of that tree about 1.5 m away at the depth of 6 m. b.g.l. the ground water will occur.

**Scientific Interpretation:**

In-Desert the date tree will grow only at that place where ground water will be available. Undewr such circumstances if two branches grow proves good ground water presence in aquifers.

**श्लोकः**

यदि भवति कर्णिकारः सितकुसुमः स्थात् पलाश वृक्षोवा ।
सव्येन तत्र हस्तद्वयेऽम्बु पुरुषद्वये भवति ॥ 59 ॥

**हिंदी:** यदि श्वेत पुष्प से युक्त कठ चम्पा का वृक्ष अथवा पलाश का वृक्ष रहे तो उस स्थान से दक्षिण दिशा में २ हाथ के बाद (4 feet or 1.5 m. b.g.l.), दो पुरुष नीचे (12 feet or 4 m. b.g.l.) के बाद जल मिलता है।

**English:** Champa tree with white flows or palash tree if available then towards south from that tree nearly 1.5 m. away, at the depth 4 m. b.g.l. ground water will be available.

**Scientific Interpretation:**

Champa/Palash are hydrophytes, they grow only such conditions where ground water is present, proves the ground water potential aquifer at deeps levels.

**श्लोकः**

यस्यामूस्मा धान्न्यां धूमो वा, तत्र वारि नरयुगले।
निर्देष्टव्या च शिरा महता तोयप्रवाहेय ।। 60 ।।

**हिंदी:** जहाँ पर गर्म पानी की भाप (गर्मी) या जहाँ पर धूम्र देख पड़े वहाँ पर दो पुरुष (12 feet or 4 m) के नीचे बड़े वेग से बहने वाली शिरा रहती है।

**English:** At places where hot water vapour is visible at that place at the depth of 4 m a good water flowing vein is visible.

**Scientific Interpretation:**

When ground water flowing vein will be visible at 4 m depth b.g.l. proves that due to aquifer water temperature – Vapour will appear.

Deep tube well water is always warm is the signal of temperature in aquifer water.

# CHAPTER 4

श्लोकः

यस्मिन् क्षेत्रोद्देशे जातं सस्यं विनाशमुपयाति ।
स्निग्धमतिपाण्डुरं वा महाशिरा नरयुगे तत्र ।। 61 ।।

**हिंदी:** जिस भूमि पर सस्य आकर के नाश हो जाती हो या जिस भूमि पर स्निग्ध अन्न बहुत हो या धान्य के पत्ते पीतवर्ण हो जावें उस भूमि पर महा शिरा विशेष जल प्रवाह से दो पुरुष नीचे बहती है।

**English:** The algae grows and dries on which land or the oily grain grows maximum and the leaves of the plant are yellow on such type of land a great water vein flows at a depth of 12 feet or nearly 4 m. b.g.l.

**Scientific Interpretation:**

Algae grows in aquatic environment and later stage dries, oily plants grows but leaves become yellow indicate the water at great deeper depth. Previously the water was present at shallow depth but later the water level goes down is the indication of such symbol.

श्लोकः

मरूदेशे भवति शिरा यथा तथातः परं प्रवस्पामि
ग्रीवा करभाणाभिव, भूतलः संस्था शिरा यान्ति ।। 62 ।।

**हिंदी:** जिस भूमि पर खूब पानी न बरसे तथा वह भूमि सूखी और रेतीली हो उसे मरुभूमि कहते हैं। यहाँ उन शिराओं के बारे में बताया गया है कि उस देश में जल की शिरा ऊंट के गर्दन के समान भूमि में जाती है, अर्थात बहुत नीची होती है।

**English:** Where there is no rain and that land is dry and sandy is known as desert. The ground water veins which flows in this area directly flow at a very deep level & no ground water situation exists in upper horizon.

## Scientific Interpretation:

In Sandy area (deserts) the ground water in the upper zone are absolutely dry and availability of ground water exist at greats depth which can be extracted by deep ground water exploration.

### श्लोकः

पूर्वोत्तरेण पीलोर्यदि वल्मीको, जलं भवति पश्चात् ।
उत्तरगमना च शिरा विज्ञेया पञ्चभिः पुरुषैः ॥ 63 ॥

**हिंदी:** जो पीलू (अखरोट मेवा) वृक्षों के ईशान में वल्मीक के मिट्टी का ढेर लगा हो तो उस पीलू वृक्ष से पश्चिम और जल 4½ हाथ (9 feet or 3 metres), पाँच पुरुष नीचे (30 feet or 10 metres b.g.l.) पर जल मिलता है।

**English:** From the tree of apricot (dry fruits) in the direction of NE – if ant's head is visible then towards west from that tree at a distance of 3 m. at a depth of 10 m. b.g.l. ground water is available.

## Scientific Interpretation:

Ant's head is a symbolic thereafter of porous formation which proves the availability of ground water.

### श्लोकः

चिन्हं दर्दुर आदौ मृत् कपिला तत्परं भवेद्धरिता ।
भवति च पुरुषेऽधोऽश्मा तस्य तलेऽम्भो विनिर्देश्यम् ॥ 64 ॥

**हिंदी:** पहले खोदने पर मेंढक के चिन्ह की मिट्टी निकलेगी उसके बाद कविल वर्ग की मिट्टी मिलेगी तदन्तर हरित वर्ण भूमि होगी और इन चिन्हों के बाद एक पत्थर के नीचे शिरा रहती है।

**English:** While digging the earth at beginning the frog color soil will appear then blue colour soil will appear after that green soil will appear. After getting all these soils under a stone ground water vein flows.

**Scientific Interpretation:**

Colour of the soil indicates that these soils are in the character of water becoming formations.

## श्लोकः

पीलोरेव प्राच्यां वल्मीकोऽतोऽर्धपञ्चमैर्हस्तैः ।

दिशि याम्यायां तोयं वक्तयं सप्तभिः पुरुषैः ॥ 65 ॥

**हिंदी:** यदि पीलू (Plum) वृक्ष से पश्चिम में वल्मीक देख पड़े तो उस पीलू वृक्ष से दक्षिण दिशा में जल 4½ हाथ के बाद (9 feet or 2.5 metres) ७ पुरुष (42 feet or 14 m. b.g.l.) प्राप्त होता है।

**English:** Towards the west of the plum tree if ant's head is visible then towards the south from that plum tree after 2.5 metre at a depth of 14 m. b.g.l. the ground water is available.

**Scientific Interpretation:**

Plum tree grow only in aquatic environment and ant's head shows the porous formation which proves the ground water availability.

## श्लोकः

प्रथमे पुरुषे भुजगः सितासितो हस्तमात्रमूर्त्रिश्च ।

दक्षिणतो वहति शिरा सक्षारं भूरि पानीयम् ॥ 66 ॥

**हिंदी:** प्रथम चिन्ह के करीब ३ मीटर के बाद सफ़ेद एवं काले वर्ण का 30 cm का सर्प दिखाई पड़ेगा बाद में दक्षिण दिशा में बहने वाली शिरा होगी जो बहुत अंशो में जल बहाया करती है।

**English:** At a depth of 3 metre b.g.l. white and black colour of snake is visible, this is the first indication in this area. After this a vein flowing towards south is visible – which distributed into various branches and proves ground water potential zone.

## Scientific Interpretation:

Presence of snake shows the more permeable soil – which will allow more water to percolate, so the formation will react as a ground water worthy zone.

**श्लोक:**

उत्तरतश्चनरीरस्पादिगृहं दासेगे जलं स्वादु ।
दशभिः पुरुषैर्जैदं पुरुषे पीतोऽत्र मण्डूकः ।। 67 ।।

**हिंदी:** जो करीर वृक्ष से उत्तर की तरफ सर्पावास हो तो उस वृक्ष से उत्तर दिशा में जल ३ मीटर के बाद २० मीटर नीचे मिलता है और वह मीठा होता है। वहाँ भी करीब २.५ मीटर खोदने के बाद पीत वर्ण मेढक निकलता है।

**English:** Towards North from Karira tree if snakes heads are visible then towards North from that tree about 3 m away from this place the ground water of good quality will be available at a depth of 20 m. b.g.l. after that digging about 2.5 m. yellow frog will appear.

## Scientific Interpretation:

Karira tree from hydrophytes family, snakes head shows porosity of soil. The rate of infiltration will be very high at this situation. After that frog of yellow appearance shows presence of ground water.

**श्लोक:**

रोहीतकस्य पश्चाद हिषास, श्चेत्रिभिः करैयभ्यि ।
द्वादश पुरुषान् खात्वा, सक्षारा पाश्चिमेन शिरा ।। 68 ।।

**हिंदी:** यदि सर्पावास वल्मीक रोहित वृक्ष से पश्चिम हो तो रोहितक वृक्ष से तीन हाथ दक्षिण के बाद (6 feet nearly 2 m) पश्चिमाभिमुख खारे जल की शिरा 12 पुरुष, 72 feet या 24 m. b.g.l. के नीचे मिलेगी।

**English:** If the snakes head is towards west from this Rohitak tree then the ground water vein flowing from west which is saline will be available at a depth of 24 m. b.g.l.

**Scientific Interpretation:**

Rohitak tree may be growing in a saline environment may be a cause for salinity in ground water.

**श्लोकः**

इन्द्रतरोर्वल्मीकः प्राग्दृश्यः पश्चिमे शिरा हस्ते ।
खात्वा चतुर्देशनरान कभिला गोधा नरे प्रथमे ।। 69 ।।

**हिंदी:** यदि वल्मीक अर्जुन वृक्ष से पूर्व दिशा में देख पड़े तो उसी इंद्र तरु से एक हाथ के बाद (2 feet or 1/2 metre), १४ पुरुष खोदने पर (84 feet या 28 metre b.g.l.) पश्चिम और नीचे बहने वाली शिरा मिलती है। जिसमें एक पुरुष (6 feet or 2 m.) खोदने पर कपिल वर्ण की गोहिया मिलेगी।

**English:** From the Arjuna tree towards east ant's head is visible then about 1/2 metre away at a depth of 28 m. b.g.l. western flowing ground water aquifer is available.

After digging about 2 m. b.g.l. at this point a dark brown colour soil will appear.

**Scientific Interpretation:**

Arjun tree is from hydrophytes family and ant's head is a another symbol shows the porous nature of soil. This proves that ground water availability will be high at this point.

**श्लोकः**

यदि वा सुवर्णनाम्न स्तरो, भवेद्वामतो भुजङ्गगृहम् ।
हस्तदये तु याम्ये पञ्चदशन रावसानेऽम्बु ।। 70 ।।

**हिंदी:** यदि अमलतास वृक्ष के बायीं और (उत्तर) सर्पावास वल्मीक देख पड़े तो उस वृक्ष से दक्षिण और जल २ मीटर के बाद ३० मीटर जमीन के नीचे मिलता है तथा वह खारा होता है।

**English:** Towards the north from Amaltash tree if snakes head is visible then towards south about 2 m. ahead at a depth of 30 m. b.g.l. the saline water is available.

**Scientific Interpretation:**

Snakes head is a symbol of porous formation of soil. Amaltash tree is growing in alkaline environment will affect on quality of water & the pH and Ec will increase in that ground water & quality will deteriorate means salinity will increase.

**श्लोकः**

क्षारं पयोऽत्र नकुलोऽर्धमानवे, तामसन्नि भश्चाश्मा
रक्ता च भवति वसुधा वहतिशिरा रक्षिणा तत्र ।। 71 ।।

**हिंदी:** प्रथम तो 1/2 पुरुष (3 feet – 1 metre) के नीचे नकुल (नेवला) तथा ताम्र वर्ण का पत्थर तदपुरांत लाल मिट्टी के बाद दक्षिण शिरा बहती है।

**English:** At first at a depth of 1 m. b.g.l. mongoose then copper colour stone then red soil after that south veins flows.

**Scientific Interpretation:**

Presence of mongoose shows the porous nature of soil. Coppery colour stone proves that moisture content is high in this soil so due to weather affect the colour of stone have been changed. Red soil is an indication of Lateritic soil proves ground water worthy zone indicates the south veins means aquifer is present at this point.

**श्लोकः**

वदरीरोहितवृक्षो सम्पृक्तौ, चेद्विनापि वल्मीकम् ।
हस्तत्रयेम्बु व्श्चात्, षोऽशभिर्मानवैर्भदति ।। 72 ।।

**हिंदी:** जिस स्थान पर बेस्का वृक्ष रोहित वृक्ष से मिले हुए हों और वहाँ पर वल्मीक हो या न हो, उन वृक्षों के पश्चिम दिशा में ३ हाथ (2 metre) के बाद १६ पुरुष (96 feet or 30 metre b.g.l.) के नीचे जल रहता है।

**English:** Ber and Rohit tree grown together and ant's head is present or not present at that place – there towards west about 2 m. away ground water will be available at a depth of 30 m. b.g.l.

**Scientific Interpretation:**

Both trees grow or one place or in symbiosis stage – indicates the soil is good and quality and ground water condition is good. Such type of characters proves the availability of ground water at deeper depth.

**श्लोकः**

सुरसं जलमादौ दक्षिणा, शिरा वहति चोत्तरेणान्या ।
पिष्टनिभः पाषाणो मृत् श्वेता वृश्चिकोऽधचरे ॥ 73 ॥

**हिंदी:** यहाँ प्रथम 1/2 पुरुष (3 feet or 1 metre) पुरुष के बाद बिच्छू दिखाई पड़ेगा तदन्तर प्रथम दक्षिण तथा द्वितीय उत्तराभिमुख मीठे पानी की शिरा बहती है।

**English:** At such places where after 1 m. depth scorpion will be visible later on at first south and later on north flowing water vein of good quality will be available.

**Scientific Interpretation:**

Scorpion will be available only in porous formations and in moisture zone so it proves that in such situation fresh ground water worthy zone will occur.

**श्लोकः**

सकरीरा चेद्वदरी त्रिभिः करैः पश्चिमेन तत्राम्भः
अष्टादशभिः पुरुषैरैशानी बहुजला च शिरा ॥ 74 ॥

**हिंदी:** जब बेर तथा करीर वृक्ष साथ-साथ दिखाई दें तो उस वृक्ष से पश्चिम दिशा में जल ३ हाथ दूरी के बाद (6 feet or 1.5 metre) १८ पुरुष, (108 feet or 35 metre) के नीचे विशेष जल संयुक्त शिरा ईशान कोण (North-East) में बहती है।

**English:** When two varieties of tree Ber and Karir grow together then towards west about 1.5 m away from those trees the water will be available towards North – East from that tree at a depth of 35 m. b.g.l.

**Scientific Interpretation:**

Ber and Karir grows together is a symbol of good quality of soil formation and the rate of percolation of ground water will be high proves availability of good aquifer and getting high discharge in these formations.

**श्लोक:**

पीलुसमेता वदरी हस्तत्रय, सम्मिते दिशि प्राचयाम् ।
विंशत्या पुरुषाणामशोष्यभम्भोऽत्र सक्षारम् ॥ 75 ॥

**हिंदी:** बेर वृक्ष यदि पीलू वृक्ष से संयुक्त हो तो उस वृक्ष से पूर्व ओर ३ हाथ (6 feet or 2 metre) की दूरी के बाद २० पुरुष के (120 feet or 40 m. b.g.l.) नीचे खारा तथा अत्यधिक जल मिलता है।

**English:** If two trees grows together Ber and Pillo then towards the east from that tree after 6 feet or 2 metre away at a depth of 120 feet (4.0 m. b.g.l.) plenty of ground water will be available of saline category.

**Scientific Interpretation:**

Growing two trees together is a symbol of good quality of soil where rate of porosity is very high which proves ground water worthy zones with very high aquifer potentiality. Pillo tree may be indicating saline nature of ground water.

श्लोकः

ककुभकरीरावेकत्र संयुतौ यत्र ककुभविल्बौ वा ।
हस्तद्वयेऽम्बु पश्चान्नरैर्भवेत् पञ्चविंशत्या ॥ 76 ॥

**हिंदी:** जहाँ पर अर्जुन एवं करीर एक साथ हों अथवा जहाँ पर ककुम और बिल्व वृक्ष एक साथ हों तो उन वृक्षों से पश्चिम दिशा में 3 1/2 metre, १ मीटर के बाद २५ पुरुष नीचे (150 feet or 50 m. b.g.l.) जल रहता है।

**English:** At those places where Arjun and Kareer trees grow together or Kakun or Arjun and bel grow together then under these circumstances towards the west from these trees about 1 m. away al a depth of 50 m. b.g.l. ground water will be available.

**Scientific Interpretation:**

Arjun and Bel act as a ground water indicator shows good soil, porous and permeable and these plants are from hydrophytes family.

श्लोकः

वल्मीकमूर्धनि यदा दूर्वा च कुशाश्च पाण्डुराः सन्ति ।
कूपो मध्ये देयो, जलमत्र नरैकविंशत्या ॥ 77 ॥

**हिंदी:** यदि वल्मीक के ऊपर की दूब तथा कुश पाण्डुर वर्ण हो जाये तो ठीक उस मध्य में हि वल्मीक के ऊपर कूप खोदने से ९ पुरुष नीचे (51 feet or 17 metre) पर जल रहता है।

**English:** If the grass and Kush above the ant's head convents in to red colour there in the centre of that above the ant's head, if the dug well will be digged there about 17 m. b.g.l. water will be available.

**Scientific Interpretation:**

Ant's head are symbolic structure in the sub-surface ground water prediction then it shows that weathering proves is at greater depth and rate of infiltration have also enriched to deeper level. Possibility of ground water is very high at these places at a shallow depth.

**श्लोकः**

भूमिः कदम्बकयुता, वल्मीके यत्र दृश्यते दूर्वा ।<br>
हस्तद्वयेन याम्ये । नरैर्जलं यञ्चविंशत्या ॥ 78 ॥

**हिंदी:** जहाँ पर भूमि कदम्ब वृक्ष से युक्त हो तथा वहाँ पर दूर्वा बाल्मीक युक्त दिखायी पड़े तो उस कदम्ब वृक्ष से दक्षिण दिशा में दो हाथ (6 feet or 2 m.) के बाद ?? गुण नीचे (150 feet or 50 metre) जल रहता है।

**English:** The land where Kadamb tree are grown which are associated with grass and ant's head then towards the south from that kadamb tree about 1.5 m. away and at the depth of 50 m. b.g.l. water will available.

**Scientific Interpretation:**

Kadamb tree is from hydrophytes family means where kadamb will grow water presence will be there. Ant's head – shows the porous formation and where Durva will grow at this point proves moisture zone. These symbols proves the presence of ground water.

**श्लोकः**

वल्मीकत्रयमध्ये रोहीतकपादपो यदा भवति ।<br>
नानावृक्षैः सहितस्त्रिभिर्जलं तत्र वक्तव्यम् ॥ 79 ॥

**हिंदी:** यदि रोहितक (गुलनार) वृक्ष के साथ अन्य वृक्ष तीन से अधिक हो और वे तीन वल्मीक के मध्य में हो तो वहाँ जल रहता है।

**English:** If associated with Rohitak (Gulnar tree) or more than three tree exist and if they exist in between the ant's head proves the ground water worthy zone.

## Scientific Interpretation:

Gulnar tree is hydrophytes and ant's head gives the signal of high porosity and maximum weathered zone proves the occurrence of ground water.

## श्लोकः

हस्तचतुष्के मध्यात् षोऽशभिश्चाङ्गलैरुदग्वारि ।
चत्वारिंशत् पुरुषान् खात्वाऽश्माऽधः शिरा भवति ॥ 80 ॥

**हिंदी:** उस रोहित वृक्ष से जो बाल्मीक के मध्य में हो उससे उत्तर दिशा में जल बहने वाली शिरा ४ हाथ १६ अंगुल (करीब २.५ मीटर) के बाद ४० पुरुष (240 feet or 80 m. approx. b.g.l.) के नीचे बहा करती है।

**English:** The ant's head is in centre where the Rohitak tree is located towards the north of that tree the water flowing vein after nearly 2.5 m away the ground water will be available at the depth of 80 m. below ground level.

## Scientific Interpretation:

Ants head and Rohitak trees are ground water indicators and the Rohitak tree is from hydrophytes family & ant's head shows porous formation proves the ground water availability.

# CHAPTER 5

श्लोकः

ग्रन्थिप्रचुरा यस्मिन् शमी भवेदुत्तरेण वल्मीकः ।
पश्चात पञ्चकरान्ते शतार्धसंख्यैर्नरैः सलिलम् ।। 81 ।।

**हिंदी:** जिस स्थान पर शमी वहुत ग्रंथि युक्त हो और वल्मीक यदि उसके उत्तर हो तो उस शमी वृक्ष से पश्चिम ५ हाथ (7.5 feet or 2.5 m) के बाद ५० पुरुष नीचे (30 feet or nearly 10 m. b.g.l.) जल रहता है।

**English:** At places where shammi trees are highly grandular and ant's head is located towards the north then towards the west from that shammitree about 2.5 m away at the depth of 10 m. b.g.l. the water will be available.

## Scientific Interpretation:

Shammi trees are really large tree and more probably comes in the category of Hydrophytes? Ant's head is another example which shows availability of ground water at deeper depth.

श्लोकः

एकस्थाः पञ्च यदा वल्मीका, मध्यमो भवेच्छ्वेतः ।
तस्मिन् शिरा प्रदिष्टा नरषष्ट्या पञ्चवर्जितया ।। 82 ।।

**हिंदी:** जब एक ही स्थान पर ५ वल्मीक हो तो उन सब वल्मीकों में से पांचवा मध्यस्थ होगा। यदि मध्यस्थ वल्मीक श्वेत हो तो उसी वल्मीक में ५५ पुरुष नीचे (330 feet, 110 m. b.g.l.) जल मिलता है।

**English:** If five ant's head are risible at one place out of which the fifth one will be centralone. If the appearance of this is white then definitely water will be available at same point at a depth of 110 m. b.g.l.

**Scientific Interpretation:**

Ant's head is a symbolic characteristic for occurrence of ground water apart from that white appearance of soil shows the presence of ground water at deeper depth.

श्लोक:

सपलाशा यत्र शमी पश्चिमभागेऽम्बु मानवैःषष्ट्या ।
अर्धनरेऽहि प्रथमं सवालुका पीतमृत् परतः ॥ 83 ॥

**Shamic + Palash Tree together**

**हिंदी:** जिस स्थान पर पलाश तथा शमी वृक्ष युक्त हों तो उस वृक्ष के पश्चिम भाग में ५ हाथ (7.5 feet, 2 metres) के बाद ६० पुरुष नीचे (360 feet or 120 m. b.g.l.) जल रहता है। प्रथम तो 1/2 पुरुष के बाद (3 feet or 1 metre) बालू सहित पीली मिट्टी होती है।

**English:** At those place where palash and shammi trees are together then towards the west from that tree about 2 metres away and at the depth of 120 m. b.g.l. the ground water will be available. At the beginning at a depth about 1 metre b.g.l. sand with yellow soil will be available.

**Scientific Interpretation:**

Palash and shammi tree are ground water indicator and at the beginning in sub surface strata availability of sand with soil will give a good infiltration for recharge of the area.

**श्लोकः**

> वल्मीकेन परिवृतः श्वेतो रोहीतको भवेद्यस्मिन ।
> पूर्वेण हस्तमात्रे सप्तत्या मानवैस्मयु ॥ 84 ॥

**हिंदी:** जहाँ पर श्वेत रोहितक वृक्ष बाल्मीक से युक्त हो उस स्थान पर रोहितक वृक्ष से पूर्व दिशा में ९ हाथ (1.5 feet or 1/2 metre) के बाद ७० पुरुष नीचे (420 feet or 140 metres below ground level) जल रहा करता है।

**English:** At those place s where white Rohitak (Kachanar) trees associated with Ant's head at that blace towards east from that tree about 1/2 metre away and at the depth of 140 m. b.g.l. the ground water is available.

**Scientific Interpretation:**

The Rohitak tree are hydrophytes and ant's head are symbolic marker features for identification of deeps ground water which proves the availability of ground water.

**श्लोकः**

> श्वेता कण्टकबहुला यत्र शमी दक्षिणेन तत्र पयः ।
> नरपञ्चकसंयुतया सप्तत्याहिर्नरार्धे च ॥ 85 ॥

**हिंदी:** जहाँ पर श्वेत वर्ण तथा अधिक काटों से संयुक शमी वृक्ष हो तो उस शमी वृक्ष से दक्षिण दिशा में ९ हाथ (1.5 feet or 1/2 metre) के बाद ७५ पुरुष के नीचे (450 feet or 150 m. b.g.l.) जल रहा करता है जिसमे 1/2 पुरुष खोदने पर (3 feet or 1 metre b.g.l.) सर्प रहता है।

**English:** A white Shammi tree which is more thorny then towards the south from that shammi tree about 1.5 metre away at the depth of 150 metres below ground level the water will be available under which after digging nearly 1 metre the snake will be visible.

## Scientific Interpretation:

The snake will be visible only at these places where moisture content will be high and shammi tree belongs to hydrophytes group indicates the presence of water.

## श्लोक:

मरुदेशे याच्चिन्हं न जाङ्गले तैर्जलं विनिर्देश्यम् ।
जम्बूवेतसपूर्वैये पुरुषास्ते मरौ द्विगुणाः ।। 86 ।।

**हिंदी:** मरुदेश में जो चिन्ह कहा गया है उसी चिन्ह से स्वल्पोदक तथा पहाड़ी क्षेत्रों में भी जल विचार करना चाहिये और जम्बू तथा बेतस वृक्ष द्वारा जो विचार कहा गया है उन्ही चिन्हों में मरुदेश में द्विगुणित नीचे जल बहना चाहिये।

**English:** As symptoms have been told regarding desert, in the same way where searcity of water or hilly terrain the availability of the water will be in the same way as it has been told for Jamun and cane trees in the same way in desertic area the water will be available at two times depth which depth have been described earlier (150 m) b.g.l.

## Scientific Interpretation:

The depth generally increases twice in comparison to alluvial terrain but the symbolic trees are same which belongs to hydrophytes group.

## श्लोक:

जम्बूस्त्रिवृता मौर्वी शिशुमारी सारिवा शिवा श्यामा ।
वीरुधयो वाशही ज्योतिष्मती गरुडवेगा च ।। 87 ।।

**हिंदी:** जामुन, त्रिवृता, मोखेल, शिशुकारी, सारिवा, हरीतक, श्यामा, पीपर, वीरुधय, बाराही, माल कमुनि, गरुच।

**English:** Jamun, Tribita, Mokhel, Shishukari, Sariba, Haritak, Shyama, Peeper, Vcerudhay, Barahi, Mal Kamuni, Garuch.

## Scientific Interpretation:

The above mentioned trees are ground water indicator.

## श्लोकः

सूकरिकमाषपर्णीव्याघ्रपदाश्चेति यद्यहेर्निलये ।

बल्लीकादुत्तरतस्तिभिः करैस्त्रिपुरुषे तोयम् ॥ 88 ॥

**Peepal Tree**

**हिंदी:** सूकरकन्द, वन उर्दी, व्याघ्रपदा आदि औषधि सर्पावास वल्मीक से युक्त हो तो उस वल्मीक से उत्तर दिशा में ३ हाथ के बाद (4.5 feet or 1.5 metre), तीन पुरुष के नीचे (18 feet or 6 metre b.g.l.) जल मिलता है।

**English:** Sukarkand, Banurdi, Vyaghrapada etc. if available with snakes ant's head then towards the north from that tree about 1.5 m away the water will be available at the depth of 6 m. b.g.l.

**Scientific Interpretation:**

Hydrophytes trees with snakes head structures are symbolic characteristics to get the ground water.

**श्लोकः**

एतदनूपे वाच्यं जाङ्गलभूमौ तु पञ्चभिः पुरुषैः ।
एतैरेव निमित्तैर्मरुदेशं सप्तभिः कथयेत् ।।89 ।।

**हिंदी:** यह उपर कहे हुये वृक्ष बहुत जलवाले देश में पाये जाते हैं और यहीं लक्षण थोड़ जल वाल देश में ५ पुरुष (30 feet or 10 m. b.g.l.) में रहता है और यही लक्षण निर्जल देश में रहने से जल ७ पुरुष नीचे (42 feet or 14 m. b.g.l.) रहता है।

**English:** As discussed earlier the availability of the water exist as such. In the water scarcity area the water is available at 10 m. b.g.l. and the same symptoms in desertic area the availability of the the water will be at 14 m. b.g.l.

**Scientific Interpretation:**

Water level goes down depending upon the hydrogeological situations.

**श्लोकः**

एकनिभा यत्र मही तृणतरुवल्मीकगुल्मपरिहीना ।
तस्यां यत्र विकारो भवति धरित्र्यां जलं तत्र ।। 90 ।।

**हिंदी:** जिस स्थान पर पृथ्वी एक समान हो और तृण वृक्ष लतादि से हीन हो और उस भूमि में यदि विकार पैदा हो जाये तो उस भूमिपर १५ पुरुष नीचे (90 feet or 30 m. b.g.l.) जल रहता है।

**English:** The earth is plain and grasses and trees are not grown, under such area the water will be available nearly at 30 m. b.g.l.

**Scientific Interpretation:**

Where the grass and trees will not grow on earth indicates that water level will go very deep under such conditions.

श्लोकः

यत्र स्निग्धा निम्ना सवालुका सानुनादिनी वा स्यात् ।
तत्रार्धपञ्चकैर्वारि मानवैः पञ्चभिर्यदि वा ।। 91 ।।

**हिंदी:** जो भूमि स्निग्ध, नीची, वालुका युक्त तथा शब्द युक्त हो और शब्द करती हो तो उस भूमि में ४ या ५ पुरुष के बाद (30 feet or 7 metre) जल रहता है।

**English:** The ground which is sticky, low sandy and creates sound then in that ground at a depth of 7 metres below ground level water will be available.

## Scientific Interpretation:

Sandy soil shows that porosity will be very high by which the rate of infiltration will increase & which will prove the availability of ground water.

श्लोकः

स्निग्धतरुणां याम्ये नरैश्चतुर्भिर्जलं प्रभूतं च ।
तरुगहनेऽपि हि विकृतो यस्यस्मात् तद्वदेव वदेत् ।। 92 ।।

**हिंदी:** जहाँ पर स्निग्ध तरु अधिकांश होते है तो उन वृक्षों से दक्षिण दिशा में जल ४ पुरुष नीचे (24 feet or 8 metre b.g.l.) रहता है। उन वृक्षों में भी जिसमे विकार (एनी वृक्ष के सदृश जल पुष्प) पैदा हो जाये उसके भी दक्षिण ४ पुरुष नीचे (24 feet or 8 metre b.g.l.) पर जल रहता है।

**English:** Where plenty of sticky trees are visible towards the south of that tree at the depth of 8 metres b.g.l. ground water will be available. Like as other trees water flows are available towards south from that tree at the depth of 8 metres b.g.l. water will be available.

## Scientific Interpretation:

Sticky trees are playing the symbolic role for ground water availability.

श्लोकः

नमते यत्र धरित्री सार्धे पुरुषेऽम्बु जाङ्गलानूपे ।
कीटा वा यत्र बिनालयेन बहवोऽम्बु तत्रापि ।। 93 ।।

**हिंदी:** जहाँ की भूमि पदाघात करने से नीची हो जाये (नीचे को चली जाये) तो चाहे बहुद्क भूमि हो या न हो परन्तु उस भूमि पर 1.5 पुरुष नीचे जल रहता है। (9 feet or 3 metre approximately)

**English:** Such grounds which are making sounds when forced or attacked by foot. The soil may be sandy or may not even then water will be seen at the depth of 3 metres b.g.l. worms lives without their houses indicate the presence of ground water.

**Scientific Interpretation:**

Soil which depress and makes sound indicates sandy soil – rate of porosity is very high & insect lives without their house shows moisture rate will be also high. Proves ground water worthy zone.

श्लोकः

उष्णा शीता च मही शीतोष्णाम्भस्त्रिभिर्नरैः सार्धैः ।
इन्द्रधनुर्भत्स्यो वा वल्मीको वा चतुर्हस्तात् ।। 94 ।।

**हिंदी:** जिस गरम प्रदेश में जहाँ की भूमि कुछ शीत हो या जहाँ पर कि शीत हि हो परन्तु कही गरम हो तो वहाँ 3 1/2 पुरुष नीचे (20 feet or nearly 7 metre b.g.l.) जल रहता है। जंगल आदि देशो में जहाँ पर इंद्र धनुषाकार या मत्स्य की तरह भूमि हो अथवा वल्मीक हो वहाँ ४ पुरुष नीचे पर (24 feet or 8 m. b.g.l.) जल रहता है।

**English:** In hot country land where earth is cool and in cold country land where the earth is hot proves the availability of water at the depth of 7 metres b.g.l., In forest area where rainbow or fish like structure exist on ground and ant's head are also visible at these places the water will be available at the depth of 8 m. b.g.l.

**Scientific Interpretation:**

The country land shows adverse nature between atmosphere and ground water which proves the symbolic ground water repository zone. Ant's head is a symbolic character for ground water appearance.

**श्लोकः**

वल्मीकानां पङ् वत्यांयद्येकोऽभ्युच्छितः शिरा तदधः ।
शुष्यति न रोहते वा सस्यं यस्यां च तत्राम्भः ।। 95 ।।

**हिंदी:** जहाँ पर अधिकांश वल्मीक हो और उनमें एक वल्मीक ऊँचा हो तो उस उच्च वल्मीक के नीचे ४ हाथ (24 feet or approximately 2 m. b.g.l.) जल की शिरा होती है। या जिस भूमि पर पैदा हुआ अनाज सूख जाता हो या बीजांकुर उत्पन्न हि न हो उस स्थान पर भी ४ हाथ (6 feet or 2 metres b.g.l.) जल रहता है केवल जाङ्गल तथा रेगिस्तान देश में।

**English:** At those places where numerous ant's head are existing out of a that one ant's head is highest then at the depth of 2 metres b.g.l. the water will be available under the same ant's head. The corn grown dries on the same land or seeds are not germinated then water will be available at the depth of 2 metres b.g.l. but this of condition is only applicable in desert and forest area

**Scientific Interpretation:**

Ant's heads are symbolic characteristic for availability of ground water. The other facts are available for forest & deserts.

**श्लोकः**

न्यग्रोधपलाशोदुम्बरैः समेतैस्त्रिभिर्जलं तदधः ।
वटपिप्पलसमवाये तद्वद्वाच्यं शिरा चोदक् ।। 96 ।।

**Bargad Tree**

**हिंदी:** पटु, पलाश, गुल्लर ये तीनों जहाँ एक साथ पाये जाये वहाँ पर ३ हाथ के नीचे (4½ feet or 1.5 m. b.g.l.) जल रहता है और वहाँ उत्तर शिरा बहती है और वट-पीतल जहाँ पर हो वहाँ भी नीचे ३ हाथ (4½ feet or 1.5 mts. b.g.l.) जल होता है।

**English:** Buett, Palash and Gullar all three are available at one place at that place water will be available at the depth of 1.5 m. b.g.l. and North vein flows at that point If buett & Peepal are conjugated together the ground water will be available at the depth of 1.5 mts b.g.l.

**Scientific Interpretation:**

Buett, Palash, Gullar and conjugation of Peepal and buett are the examples for ground water availability – because these plants will grow only in such places where plenty of water will be available.

**श्लोकः**

आग्नेये यदि कोणे ग्रामस्य पुरस्य वा भवत् कूपः ।
नित्यं स करोति भयं दाहं च समानुषं प्रायः ।। 97 ।।

**हिंदी:** जो कूप ग्राम के अग्नेय कोण में हो तो उस कूप से ग्राम वालों का नित्य भय रहेगा और प्रायः मनुष्य की उस कूप के कारण अग्निमय रहता है।

**English:** The will located in south-east that well creates terror every day and generally it is observed that dug will create a terror due to fire.

**Scientific Interpretation:**

South-east corner indicates the place of fire where existence of water will not be possible.

**श्लोकः**

नैर्ऋतकोणे बालक्षयं च वनिताभयं च वायव्ये ।
दिक्त्रयमेतत्तयक्तवा शेषासु शुभावहाः कूपाः ॥ 98 ॥

**हिंदी:** यदि कूप ग्राम के नैर्ऋत्य कोण में हो तो बालको को क्षय करता है, वायव्य कोण में कूप हो तो स्त्रियों की भय करता है, अतएव कूप तीन कोणों को छोड़ सब दिशा में शुभप्रद है।

**English:** If dug well is located towards Nereqtya Kone (SE) that gives loss to the children and if located towards Vayabya Kone (NW) – gives terror to the ladies. So after leaving these three directions the dug wells are fruit full in all directions.

**Scientific Interpretation:**

**श्लोकः**

मारस्वतेन मुनिना दकार्गलं यत् कृतं तदवलोक्य ।
आर्याभिः कृतमेतद् वृत्तैरपि मानवं वक्ष्ये ॥ 99 ॥

**हिंदी:** वाराह मिहिर कहते है कि यह दकार्गल सारस्वत मुनि की रचना अनुसार आर्या छंद में कहा और अब मनुकृत दकार्गल को भिन्न-भिन्न छन्दों में कहूँगा।

**English:** The Varah mihir (stated that this Dakargal – as per the geometry of Saraswat Muni have been stated Now it will be stated by Manu as he speaked in different chhandas.

**Scientific Interpretation:**

Another view which has been given by manu have been stated in these procedures. These methods should also be identified prior to select any point from ground water point of view.

**श्लोकः**

स्निग्धा यतः पादपगुल्मवल्लयो निश्छिद्रपत्राश्च ततः शिरास्ति ।
यद्यक्षुरोशीरकुलाः सगुण्ड्राः काशाः कुशा वा नलिका नलो वा ।। 100 ।।

**हिंदी:** जिस स्थान पर शाखा के समूह लता स्निग्ध वृक्ष तथा छिद्र रहित पत्ते हो उस स्थान पर ३ पुरुष के नीचे (18 feet – 6 metre b.g.l.) शिरा रहती है।

**English:** At those places where trees are grown in a group and Lata – (Creepers) and tree are sticky and no hole in leaves at those places the water will be available at the depth of 6 metres b.g.l.

**Scientific Interpretation:**

Growing of trees in a group is a good symbol for ground water availability and when the trees and sticky & no hole in leaves shows plenty of water.

# CHAPTER 6

**श्लोकः**

खर्जूरजम्ब्वर्जुनवेतसाः स्युः क्षीरान्विता वा द्रुयगुल्मवल्ल्यः ।
छत्रेभनागाः शतपत्रनीपाः स्युर्नक्तमालाश्च ससिन्दुवाराः ॥ 101 ॥

**हिंदी:** खजूर, जामुन, अर्जुन, बेत, क्षीर, वृक्ष गुल्मलताइन (रतालू) नाग (नागकेशर), शतपत्र (कमल) नीम वृक्ष, विशेष नक्तमाल, सिंदुवार (सेधुवार) वृक्ष) से युक्त हो।

**English:** Khajoor, Jamun, Arjun, Cane Ksheer, Ratallu, Nag, Lotus, Neem, Naktamal, Sindhuwar trees are available.

**Scientific Interpretation:**

The above mentioned trees grows only at those places where water is available.

**श्लोकः**

विभीतको वा मदयन्तिका वा यत्रास्ति तस्मिन् पुरुषत्रयेम्भः ।
स्यात् पर्वतस्योपरि पर्वतोऽन्यस्तत्रापि मूलेपुरुषत्रयेऽभः ॥ 102 ॥

**हिंदी:** बहेणा दमयन्ती जिस जगह हो वहाँ भी ३ पुरुष के नीचे जल रहता है। जहाँ पर्वत के उपर पर्वत रहता है वहाँ पर्वत के मूल में तीन पुरुष नीचे जल रहता है।

**English:** Bahere and Damyanti are grown at those places the water will be available at 18 ft b.g.l. 6 mts b.g.l. Where mountains are available attach with mountains then at the Joint plane water will be available at 18 ft – 6 metre b.g.l.

## Scientific Interpretation:

Bahera and Damayanti are ground water indicator plants which proves the ground water availability. The mountain plains at a Joint Plane – a weak plane develops which allows water to percolate and we get ground water at that point.

## श्लोक:

या मौञ्जिकैः काशकुशैश्च युक्ता नीला च मृद्यत्र सशर्करा च ।
तस्यां प्रभूतं सुरसं च तोयं कृष्णाशता गन च रक्तगृह्य ।। 103 ।।

**हिंदी:** जो भूमि मूंज काश कुशादि से संयुक्त हो और नील वर्ण की मिट्टी कणों से युक्त हो वहाँ उस भूमि के नीचे बहुत मीठा जल रहता है। जिस भूमि की मिट्टी काली या लाल हो वहा भी स्वादिष्ट जल निकलता है।

**English:** The soil which is associated with Munj, Kush etc and the soil grains are of blue colour there the excellent quality of ground water is available. The soil either black or red the fresh water also appears at those places.

## Scientific Interpretation:

Kush etc are a symbolic structure of hydrophytes and the soil colours blue, black & red shows rate of infiltration is high – gives a better scope for ground water availability.

## श्लोक:

सशर्करा ताम्रमही कषायं क्षारं धरित्री कमिला करोति ।
आपाण्डुरायां लवणं प्रदिष्टं मृष्टं पयो नीलवसुन्धरायाम् ।। 104 ।।

**हिंदी:** जो ताम्र वर्ण की भूमि कंकणो से युक्त हो वहाँ का जल कषाय होता है और यदि कपिल वर्ण भी हो तो क्षार जल देती है। पाण्डुर (लाल सफ़ेद) वर्ण की भूमि हो तो नमक के स्वाद का जल होता है। तथा नीली भूमि हो तो मीठा जल देती है।

**English:** The soil coppery in colour associated with Kankars the water will be bitter there. If the soil is whitish in appearance then yields saline water. If it is reddish white then yield salty water and if the soil is blue then gives fresh water.

**Scientific Interpretation:**

Different taste of water appears where it comes in contracts of different type of soils. For eg. Coppery soil gives bitter taste – colour shows the higher grade of saltification. Blue soil gives the fresh quality.

**श्लोकः**

शाकाश्वकर्णार्जुनविल्वसर्जाः श्रीपण्ययरिष्टाधवशिंशपाश्च ।
छिद्रैश्च पत्रैर्द्रुमगुल्मवल्ल्यो रुक्षाश्च दूरेऽम्बु निवेदयन्ति ॥ 105 ॥

**हिंदी:** यदि छिद्र रहित पत्तों से युक्त, गुल्म तथा वल्ली के सहित शाक, अश्वकर्ण, अर्जुन, विल्व सर्ज श्रीपर्णी अरिष्ट धव शिशया के वृक्ष हों तो वहाँ बहुत दूर नीचे जल रहता है।

**English:** If the above said tree which are not porous. If available in that area proves ground water worthy and also indicate that water will be available at greater depth.

**Scientific Interpretation:**

The above mentioned tree shows the presence of water at greater depth.

**श्लोकः**

सूर्याग्निभस्मोष्ट्रखरानुवर्णा या निर्जला सा वसुधा प्रदिष्टा ।
रक्तांकुराः क्षीरयुताः करीरा रक्ता धराचेज्जलभश्मनोऽधः ॥ 106 ॥

**हिंदी:** जो भूमि सूर्य, अग्नि भस्म, उष्ट्र तथा गदहे की रंग की हो वह निर्जला होती है। जहाँ पर करीर वृक्ष रक्तवर्ण अंकुर तथा दूधवर्ण से युक्त हो वहाँ पर पत्थर के नीचे जल रहता है।

**English:** The soil is like as sun, fire ashes, camel & donkey like colour – shows no water. At those places where karrer tree appearance reddish in colour and milky white in colour at those places ground water is available below stones.

**Scientific Interpretation:**

The colour of the soil is red, camel or donkey in colour – shows no moisture in the soil which proves absence of water.

**श्लोक:**

वैदूर्यमुद्गाम्बुदमेचकाभा पाकोन्मुखोदुम्बरसन्निभा वा ।
णुणाज्ञानाम कपिलाथवा या ञया शिला भूरेसमीपतोया ॥ 107 ॥

**हिंदी:** यदि पत्थर वैदूर्य मणि के तुल्य मुद्ग (मूंग) कृष्ण वर्ण हो तथा पके गूलर फल के समान लाल हो तथा जिस पत्थर के तोड़ने से सुर्मा के वर्ण का चूर्ण हो या सफ़ेद हो, उस पत्थर के समीप बहुत जल है ऐसा जानो।

**English:** If the stone is like as precious stone and its appearance is like as a shining black of riped gular fruit – red – appearance and if it is hammering then black or white particles appears from that then by the side of the stone plenty of water is available.

**Scientific Interpretation:**

Shankargarh sand stone in Bundelkhand is a best example in Indian conditions. We get plenty of water after exploring this area.

**श्लोक:**

पारावतक्षौद्रधृतोपमा या क्षौमस्य वस्त्रस्य च तुल्यवर्णा ।
या सोमवल्ल्याश्च समानरुपा साप्याशु तोयं कुरुतेऽक्षयं च ॥ 108 ॥

**हिंदी:** जो भूमि कबूतर के समान हो, मधु या घृत के सदृश हो या रेशमी वस्त्र के तुल्य हो अथवा जो भूमि सोम वल्ली (लता) के सदृश हो तो वह बहुत शीघ्र अक्षय जल को करती है।

**English:** The earth – (soil) whose appearance is like as a pigeon, Honey or Ghee or like as a silky clothes or the appearance of soil like as a creepers that gives plenty of ground water.

## Scientific Interpretation:

The interpretation have been made based on the colours of the soil for example pigeon, Honey or Ghee like colour. These colours shows that the moisture content in the soil is high – by which water percolates to sub surface and yields good quantum of water.

## श्लोकः

तामैः समेता पृबतैर्बिचित्रैरापाण्डुभस्मोष्ट्रखरानुरुपा ।
भृङ्गोपमांगुष्ठिकपुष्पका वा सूर्याग्निवर्णा च शिला वितोया ॥ 109 ॥

**हिंदी:** जो पत्थर ताम्र वर्ण के पृषत (तिलक) से युक्त हो या नाना वर्ण के चिन्हों से अंकित हो या जिस शिला का पांडुवर्ण अथवा भस्म ऊँट या गधा के तुल्य रंग हो अथवा भ्रमर के समान हो या अंगुष्ठ के वृक्ष के फूल में तुल्य हो या सूर्य अग्नि के सदृश हो वहाँ शिरा जल से रहित होती है।

**English:** Stone Coppery in colour and different types of symbols are present on that red colour rock or colour is like a donkey or camel or beatel - the water is not available at that place.

## Scientific Interpretation:

Based on the colour of the stone the water presence have been described so colours like as coppery, donkey colour, as described not shows presence of water.

## श्लोकः

चन्द्रातपस्फटिकमौक्तिकहैमरुपा याश्चेन्द्रनीलमणिहिंगुलुकाञ्जनाभाः ।
सूर्योदयांशुहरितालनिभाश्च याः स्युस्ता शोभना मुनिवचोऽत्र च
वृत्तमेतत् ॥110॥

**हिंदी:** जो शिला चन्द्र कान्ति समान स्फटिक तुल्य मोती के सदृश तथा सुवर्ण या इंद्र नीलमणि वा हिंगुलक (लाल) वर्ण के तुल्य हो या अंजन सदृश्य कांतिवाला या सूर्योदय के समय सूर्य बिम्ब के तुल्य कांतिवाला या हरिलाल के समान हो वह उत्तम है, ये सब श्लोक मुनिवचन है।

**English:** The stone like as sapphire and like as pearl and reddish in appearance and shining black in appearance and at the time of sun shining that shines like as sun these are the positive symbols as stated in slokas.

**Scientific Interpretation:**

The above stated symbols are symbolic characteristics for getting better environment/ground water.

## श्लोकः

एता ह्यभेद्याश्च शिलाः शिवाश्च यक्षैश्च नागैश्च सदाभिजुष्टाः ।
येषां च राष्ट्रेषु भवन्ति राज्ञां तेषामवृष्टिर्न भवेत् कदाचित् ।। 111 ।।

**हिंदी:** सब पूर्वोक्त शिला अभेद्य है। क्योंकि ये कल्याणकारी हैं। यक्ष नाग सेकितये शिला जिस राजा के राज्य में होती हैं उसमें कभी भी अवर्षण नहीं होता।

**English:** All stones are unbrakable because they always do the best for the welfare of the public. By processed with Yaksha Naga these stones where kept always there will be the rain in the area & there will be no famine.

**Scientific Interpretation:**

The scientific role at this state cannot be predicted but it can be assumed that such colour stones if will be visible in the area make the area more prosperous in the field of ground water.

## श्लोकः

भेदं यदा नैति शिला तदानीं पलाशकष्ठैः सह तिन्दुकानाम् ।
प्रज्वालयित्वानलमग्निवर्णां सुधाम्बुसिक्ता प्रविदारमेति ।। 112 ।।

**हिंदी:** यदि शिला तोड़ने का विहार हो और वह न टूटे तो तत्काल तेंदू के और पलाश के काष्ठको उसके उपर अग्नि जला दें जब शिला लाल हो जाये तो उसको चूना के पानी में ठंडी करें तो शिला फूटती है।

**English:** For breaking of the stone fire the woods of plash and tendu and when the stone (slab) becomes red then apply lime water or that it will affect immediately.

**Scientific Interpretation:**

The styles by burning fine wood give a new direction for fracturing in hard rocks for ground water infiltration.

**श्लोकः**

तोयं श्रितं मोक्षकभस्मना वा यत् सप्तकृत्वः परिषेचनं तत् ।
कार्यं शरक्षारयुतं शिलायाः प्रस्फोटनं वाह्नवितापितायाः ॥ 113 ॥

**हिंदी:** मोक्षक (मरुतक जमीरी निम्बू) वृक्ष के भस्म में गरम किया जल शर नामक तृण के भस्म को मिलाकर गर्म की हुई शिला को सात बार सींचे पुनः गरम करने और सींचे तो शीला फूटती है।

**English:** The ash of lemon tree add with hot water and the ash of shir – grass is sprinkled seven times then repeat this process then definitely the slab will break.

**Scientific Interpretation:**

This is the oldest scientific approach to break the slabs/boulders. The chemical constituents which reacts for breaking the slabs – should be identified and this should be correlated with the present scientific approach.

**श्लोकः**

तत्र्ककाञ्चिकसुराः सकुलत्था योजितानि बदराणि च तस्मिन् ।
सप्तरात्रमुषितान्यभितप्तां दारयन्ति हि शिलां परिषेकैः ॥ 114 ॥

**हिंदी:** मठा, काजी, मध्य, कुलित्थ उन सबको एकत्र करके उसमे बेर डाले। सात रात्रि एकत्र रखकर उसमे गरम शिला को ठंडा करने से शिला फूटती है।

**English:** Butter milk, Kawjee, wine, Kulith all added with strow berry. Continuously 7 nights the warm slabs should be kept to cool in it – Then the slab will break.

**Scientific Interpretation:**

This is ancient mythology. Scientist has to work in this field to know the fact to work in the field of hydrofracturing.

**श्लोकः**

नैम्पं पत्रं त्पक् च नालं तिलानां सापामार्ग तिन्दुक स्याद् गुडूची ।
गोमूत्रेण स्रावितः क्षार एषां षट्कृत्बोऽतस्तापितो भिद्यतेऽश्मा ॥ 115 ॥

**हिंदी:** नीम के पत्र और क्षाल, तिलनाल चिचिणा, तेंदुआ और गुरूच इनका भस्म गोमूत्र से युग्म करें उसमे गर्म किया पाषाण छः बार पुनः गरम करके सींचने से फूटता है।

**English:** Neem leaves and barks, Tilnal Chichira, Tendua and Guruch the ash of these plants added with cow's urine and it six time this urine water is making hot and slab in dips to it. By this repetition process the slab will break easily.

**Scientific Interpretation:**

The chemical qualities of the above mentioned plants should be identified and it should be correlated with scientific approach.

**श्लोकः**

आर्क पयो हुडुविषाणमषीसमेतं पारावताखुशकृता च युतः प्रलेपः ।
टङ्कस्य तैलमथितस्य ततोऽस्य पानं पश्चाच्छितस्य न शिलासु
भदेद्विघातः ॥ 116 ॥

**हिंदी:** मटर के दूध में मेढ़ा के सींग की राख मिला के फिर उसमें कपूर और चूहा के विष्ठा को मिला के उसका लेप जिस टांगी पर तेल मला हुआ होआ उस पर लेप करें फिर उसको पानी से धोवें और फिर घिस

के चोंक भरे तो पत्थर तोड़ने के समय टांगी नहीं टूटेगी। अतः तलवार आदि पर नियत क्रिया करने से तलवार आदि नहीं टूटेगी।

**English:** In the milk of Aaak. The ash of the horn of mevdha with camphor is add with the excreta of rat – apply on the nail or sword – it will never break.

## Scientific Interpretation:

It should be identified with the proper chemical constituents for future research.

## श्लोकः

क्षारे कदल्या मथितेन युक्ते दिनोषिते पायितमायसं यत् ।
सम्यक् शितं चाश्मनि नैति भङ्ग न चान्यलोहेष्वपि तस्य कौण्ठ्यम् ॥ 117 ॥

**हिंदी:** केला क्षार (राख) मठा से युक्त काके अहो रात्र एकत्र रक्खें फिर उस पानी को जिस लोहे के शास्त्र (तलवार, बरछी, छुरी, टामी आदि) को खूब पिलावें अर्थात खूब भिगोवें तो वह पत्थर पर तथा अन्य लोहे पर कुंठित न होगा।

**English:** Banana stem ash dipped in butter milk for a whole night then the instrument should be dipped in that for hours together then that instrument will never bent after prolonged use.

## Scientific Interpretation:

The chemical constituents what are present in banana ash and how it affects and what reactions takes place with milk to give the soliditication to the instrument a research should be made at these points.

## श्लोकः

पाली प्रागपरायताम्बु सुचिरं धत्ते न याम्योत्तरा ।
कल्लोलैरवदारमेति मरुता सा प्रायशः प्रेरितैः ।
तां चदिच्छति सारदारुभिरपां संपातमावारयेतः ।
पाषाणादिभिरेव वा प्रतिचयं क्षुण्ण द्विपाश्वादिभिः ॥ 118 ॥

**हिंदी:** पूर्व, पश्चिम पाली (कूप, पोखरा, वावली) बहुत दिन जल धारण करते हैं, दक्षिण उत्तर दिशा में बहुत दिन नहीं रहते हैं क्योंकि वायु प्रेरित तरंगों से उनका विशेष विदारण होता है। ऐसे कूप को जो बनाने की इच्छा करे वह जिधर जल तरंग जाते हो उधर दृढ काष्ठ से या पत्थर से अथवा ईंट मूंज से जल पतन स्थल को बनावें। द्विपद (अश्व, गज, ऊंट, बैल आदि के समूह से उस स्थान को मर्दन करावें तो मजबूत होता है।

**English:** If the dng wells/Ponds are constructed towards east/west. They get water continuously at the same time when they are in North – South direction they dry. The well should be constructed in the same direction in which direction wind if flows.

## Scientific Interpretation:

It is related with the earth magnetic field. We get water in East-West instead of NS in which direction we do not get water.

## श्लोक:

ककुभवटाम्रप्लक्षकदम्बैः सनिचुलजम्बूवेतसनीपैः ।
कुरुवकतालाशोकमधूकैर्बकुलविमिश्रै श्चावृत्तीरा ।। 119 ।।

**हिंदी:** अर्जुन, वट, आम, पाकर, इमली, कदम्ब, बेल, नीम, जामुन, कट सरई, ताड़, अशोक, अद्रुक, मौसरी यह वृक्ष जलाशय के तट पर लगाना है।

**English:** Arjun, Butt, Mango, Paker, Metarind, Kadams, Cane, Neem, Jamun, Katsara, Tar, Ashok, Mahva, Maushari – all such types of tree should be grown along the rivers, tanks or ponds etc.

## Scientific Interpretation:

The above mentioned trees belongs to hydrophytes groups which proves that only in shallow water level conditions these plants can grow and act as a artificial recharge by plantation.

श्लोकः

द्वारं च नैर्वाहिकमेकदेशे कार्य शिलासंज्चितवारिमार्गम् ।
कोशस्थितं निर्विवरं कपाठं कृत्वा ततः पांसुभिराकत्तम् ।। 120 ।।

**हिंदी:** बावली, तालाब या बड़े कूप आदि में पानी आने के लिए पत्थर से
सदृढ़ नाली नहर बनवायें, उसके उपर रक्षार्थ द्विद रहित केवाण लगाना
और मृत्रिका से आच्छादित करना चाहिए।

**English:** Ponds, tanks, large dia dug wells etc. to get water the lined canal should be constructed and above that non porous cover should be part by which the dirt should not percolate in to it.

**Scientific Interpretation:**

Ponds – tanks – water should be filled by running canal by which continuous supply can be made for longer period.

श्लोकः

अञ्जन मुस्तोशीरः सराज कोशात कामलक चूर्णैः ।
कनक फल समायुतै र्योगः कूपे प्रदातव्यः ।। 121 ।।

**Impact of Sewage Water on Ground Water Pollution**

**हिंदी:** अंजन (कुटकी काला कपास), नागरमोथा, खस, बन तोरई या बड़ी तोरई इनका चूर्ण और कनक फल (निर्मला का बीज) को एकत्र करके कूप में छोड़े।

**English:** Black Cotton, Nagarmotha, Khus, forest torai & powder of it and Kanak fruit should be collected and dip in the well.

## Scientific Interpretation:

The above said plant have some medicinal value for purification of drinking water.

## श्लोकः

कलुषं कटुकं लवणं विरसं सलिलं यदि वा शुभगन्धि भवेत् ।
तदनेन भवत्यमलं सुरसं सुगन्धि गुणैर परैश्च युतम् ।। 122 ।।

**हिंदी:** उससे मैला कटुक्षार नीरस अशुभगंधवाला भी जल स्वन्छ, सुरस, सुगन्धित, इस गुण से और अपर गुण से भी युक्त होता है।

**English:** Dirt, hardness, tasteness, unpleasant smell, ground water, convert in to fresh water, tasty and without any odour. By this process it may convert better than this also.

## Scientific Interpretation:

Some chemical constituents are present which purify the polluted water in a good quality water.

Definitely this is a research point which process great significance for purification of drinking water by natural process.